Lecture Notes in Mathematics

Edited by A. Dold and B. Eckmann
Subseries: Harvard/MIT
Adviser: G. Sacks

T0226022

521

Greg Cherlin

Model Theoretic Algebra
Selected Topics

Springer-Verlag
Berlin · Heidelberg · New York 1976

Author
Greg Cherlin
Department of Mathematics
Rutgers University
New Brunswick
New Jersey 08903/USA

Library of Congress Cataloging in Publication Data

Cherlin, Greg, 1948-
Model theoretic algebra.

(Lecture notes in mathematics ; 521)
Bibliography: p.
Includes indexes.
1. Model theory. 2. Algebra. I. Title.
II. Series: Lecture notes in mathematics (Berlin)
QA3.L28 no. 521 [QA9.7] 510'.8s [512'.02] 76-15388

AMS Subject Classifications (1970): 02H15

ISBN 3-540-07696-4 Springer-Verlag Berlin · Heidelberg · New York
ISBN 0-387-07696-4 Springer-Verlag New York · Heidelberg · Berlin

Printed in Germany

Printing and binding: Beltz Offsetdruck, Hemsbach/Bergstr.

Table of Contents

Introduction

No attempt has been made in this volume to give a balanced survey of model theoretic algebra as a whole; on the contrary a few main themes have been emphasized, most of which were championed by the late Professor Abraham Robinson. In particular we lay great stress on the role played by transfer theorems and existentially complete structures in algebra (these notions are explained and illustrated in Chapters I-II).

The study of "model theoretic" algebra naturally requires a certain knowledge of the leading ideas and methods of both subjects. The necessary model theoretic background is sketched in a preliminary Chapter 0, whereas useful algebraic machinery is developed only as the need arises. Succeeding chapters have been kept somewhat independent of one another, with the major exception of the pivotal Chapter III §1, which is motivated by material in Chapters I-II, and is essential for much of Chapters IV-V. (Not surprisingly, the results on modules in Chapter V are intimately related to the more precise results of Chapter VI on abelian groups.)

In more detail, the structure of these lectures is as follows. Chapter I is devoted to examples of the use of transfer theorems in algebra. We then apply the Ax-Kochen-Ershov transfer theorem for valued fields to Artin's conjecture and other problems in the theory of Hensel fields. Our third chapter introduces the study of general existentially complete structures, a topic which is suggested naturally by the interpretation of Hilbert's Nullstellensatz as a transfer theorem for fields (Chapter I §2). Our discussion of the general theory of existentially complete structures is illustrated by the more concrete considerations of Chapters IV-V (dealing respectively with division rings and modules). We then treat the model theory of abelian groups, following Eklof and Fisher, and close in Chapter VII with a rather different topic, more in the spirit of

[43,49]: Macintyre's analysis of $\aleph_1$-categorical fields.

Exercises have been provided for each chapter, with no indication of their relative difficulty (which varies greatly).

Acknowledgements

The notes following each chapter have been included primarily as clues for the reader who may wish to investigate the published literature; it would be a grave (and improbable) error to interpret them as an adequate expression of my indebtedness to my colleagues. In the same spirit, the bibliography has been kept extremely compact, and is restricted to works actually cited in the text.

These lectures are based on the notes for a graduate course I gave at M.I.T. in the spring of 1974, and again at the University of Heidelberg in the spring of 1975. I am deeply grateful to both universities, and specifically to Professors G. Sacks and G. Müller, for their support during this period.

I would also like to acknowledge the intellectual and personal debt I owe to the late Professor Abraham Robinson. In ways too numerous to detail he exerted a profound and continuous influence on the development of the subject treated here.

New Brunswick, N.J.
Dec. 4, 1975

0. Basic Model Theory

Introduction. For the reader's convenience we summarize certain basic notions of model theory which are treated at length in [11,36,43]. Foremost among these is the notion of a first order sentence (§1).

The most important theorem reviewed in this chapter is the Compactness Theorem (§1). This is most efficiently exploited via saturated models (§2).

§1. First order languages. First order sentences.

Structures and language are treated carefully in [11,36,43]. We will review the salient points.

A mathematical structure $\mathfrak{A}$ (also called a relational system or model) consists of a set A on which various functions f_i and relations R_j are defined; in addition various elements c_k of S may be distinguished. Thus a structure is a 4-tuple: $\mathfrak{A} = \langle A;\{f_i\},\{R_j\},\{c_k\}\rangle$. For example an ordered abelian group A will be equipped with a binary operation $t:A\times A\to A$, an ordering $<$ (a binary relation on A), and a distinguished element 0 (the identity):

$\mathfrak{A} = \langle A,\{+\},\{<\},\{0\}\rangle$.

If $\mathfrak{A} = \langle A;\{f_i\},\{R_j\},\{C_k\}\rangle$ is a structure, where f_i is a function of m_i arguments and R_j is a relation of n_j-tuples of elements of A, we call m_i or n_j the rank of f_i or R_j. If $\mathfrak{A}' = \langle A',\{f_i'\},\{R_j'\},\{c_k'\}\rangle$ is a second structure in which rank f_i' = rank f_i, rank R_j' = rank R_j for all i, j, then a map $h:A \xrightarrow{1-1} A'$ will be called an injection (monomorphism) of $\mathfrak{A}$ into $\mathfrak{A}'$ if it respects the functions, relations, and distinguished elements of $\mathfrak{A}$ and $\mathfrak{A}'$. In more detail we assume:

1. $h:A \to A'$ is one-one
2. $h(f_i(a_1,\ldots,a_{n_i})) = f_i'(ha_1,\ldots,ha_{n_i})$

3. $R_j a_1,\ldots,a_{n_j}$ holds if $R_j' ha_1,\ldots,ha_{n_j}$ holds.
4. $hc_k = c_k'$.

Surjective injections are called bijections or isomorphisms. As we will make comparatively little use of general homomorphisms, we do not pause to define them here.

We now turn from the mathematical structures to a consideration of formal statements about mathematical structures. Consider the following simple statements in the language of ordered abelian groups:

1. $\forall x \forall y (x+y=y+x)$,
2. $\forall x \exists y (y+y=x)$,
3. $\exists x (x>0 \& \forall y (y>0 \Rightarrow [y>x \text{ or } y=x]))$.

The first statement is of course true in all abelian groups, while the second and third statements may be either true or false, depending on the group in question.

In formulating statements 1-3, we made use of:

a. Variables x, y, varying over the elements of our domain A.
b. Quantifiers $\forall$, $\exists$ (for all, there exists).
c. Relation symbols $>$, $=$; every ordered abelian group is of course equipped with a definite ordering $>$, and in addition every set whatsoever carries a canonical equality relation $=$.
d. A function symbol $+$: every ordered abelian group is equipped with a binary function $+$.
e. A constant 0 denoting the distinguished elements
f. Logical connectives (and, or, implies). Other connectives one might use are $\neg$, $\Leftrightarrow$ (read "not," "iff).

Punctuation: [ ] ( ) etc.

Mathematical statements which can be written down using only the symbols listed above (variables, quantifiers, relation, function symbols, constants, logical connectives, punctuation) are called first-order statements (in the language of abelian groups). Thus statements 1-3 are first order. A typical statement which is not first order is:

$$\forall x \exists n\, [\underbrace{1+\ldots+1}_{n \text{ times}} > x].$$

This statement offends us in various ways. For example, we see the variable n which represents a variable integer rather than an arbitrary member of our group.

Let us briefly consider ordered fields from this point of view. An ordered field $\mathfrak{F}$ is a 4-tuple:

$$\mathfrak{F} = \langle F\ ; \{+,\cdot\},\ \{>\},\ \{0,1\} \rangle.$$

The following statements are first order and are true in all ordered fields:

1. $\forall x \exists y [x=0 \text{ or } x\cdot y=1]$.
2. $\forall x [x+1>x]$.

The following statement is first order, and frequently true:

3. $\exists x (x>0\ \&\ \forall y [y^2 \neq x])$.

The following statement is occasionally true, but is not first order:

4. $\forall S \subseteq F \forall T \subseteq F$ [If $\forall x \varepsilon S\ \forall y \varepsilon T (x \leq y)$ and $S \neq \phi$ and $T \neq \phi$ then $\exists z\ \forall x \varepsilon S\ \forall y \varepsilon T\ (x \leq z \leq y)$].

(The problem with statement 4 is its use of set-theoretical concepts.)

Naturally the language of ordered fields is more expressive than the language of abelian groups. We have an additional function symbol $\cdot$ at our disposal, as well as the extra constant 1. In general, a first order alphabet consists of a supply of variables, quantifiers, Logical connectives, and punctuation together with a definite supply of function symbols, relation symbols, and constants. Given a particular first order alphabet, the corresponding first order language consists of all meaningful first order statements which can be written using the given alphabet. Thus the first order alphabet appropriate to ordered fields is larger than the first order alphabet appropriate to ordered abelian groups, and the first order language of ordered fields is, correspondingly, more expressive. In general, we specify a first order language by describing the supply of function symbols, relation symbols, and constants available (to be used in conjunction with variables, quantifiers, logical connectives, punctuation, and the equality symbol). Thus to describe the language of abelian groups we note the availability of a binary function symbol $+$, a binary relation symbol $>$, and a constant symbol 0.

Definition 1. A theory T is an arbitrary set of first order sentences (in a given first order language L). A structure $\mathfrak{A}$ is a model of T (or: satisfies T) iff every statement in T is true of $\mathfrak{A}$. T is consistent iff T has at least one model. T is complete (relative to L) iff for every sentence s of L, either s or its negation $\neg s$ is in T.

Examples 2. The language L of group theory contains a binary relation symbol $\cdot$ and a constant 1. The axioms of group theory are customarily given as three first order assertions:

1. $\forall x \ \forall y \forall z \ [(x\cdot y)\cdot z = x\cdot(y\cdot z)]$
2. $\forall x \ [x\cdot 1 = 1\cdot x = x]$
3. $\forall x \ \exists y \ [xy = yx = 1]$.

If we let T be the theory consisting of these three axioms, then it is consistent (since there are groups) and incomplete (neither the axiom of commutativity nor its negation are in T). T has finite models and infinite models.

The following theorem is fundamental.

Theorem 3 (Compactness). A theory T is consistent iff every finite subset of T is consistent.

The power of this theorem will become evident in the course of this volume (see e.g. Chapter I). The proof goes as follows. The notion of consistency given in definition 1.1 is related to a second, ostensibly weaker notion: call a theory logically consistent iff no contradiction can be deduced from it (see the references cited for a discussion of the rules of inference in formal logic). It is evident upon brief reflection that a theory T is logically consistent iff every finite subset of T is logically consistent. On the other hand it is possible to show that consistency and logical consistency are equivalent for first order theories, and theorem 3 follows.

As a corollary to the compactness theorem, we have:

Theorem 4. Let s be a first order statement in the language of fields. Assume that there are arbitrarily large finite primes p such that s holds in some algebraically closed field of characteristic p. Then s holds in some algebraically closed field of characteristic 0.

Proof: Let T be the theory of algebraically closed fields, and let X_p be the axion: "The characteristic is p." Let $T' = T \cup \{\neg X_p:$ all primes $p\} \cup \{s\}$. If T' is consistent, then any model of it will be the desired algebraically closed field of characteristic 0. If T' is inconsistent, the compactness theorem tells us that some finite subset of T' is already inconsistent. But if $T \cup \{\neg X_p : p<N\} \cup \{s\}$ is inconsistent, then no algebraically closed field of characteristic greater than N satisfies s, a contradiction. $\dashv$

We conclude this section with a look at one source of complete and consistent theories.

Definition 5. Let $\mathcal{A}$ be a structure, and let L be an appropriate first order language (L will have function symbols, relation symbols, and constants denoting the functions, relations, and distinguished elements of $\mathcal{A}$). Then Th($\mathcal{A}$) (the theory of $\mathcal{A}$) is defined to be the set of all first order statements (in the language L) which happen to be true of $\mathcal{A}$. Th($\mathcal{A}$) is evidently complete and consistent. Given a second structure $\mathcal{B}$, we will say that $\mathcal{B}$ is elementarily equivalent to $\mathcal{B}$ (in symbols: $\mathcal{A} \equiv \mathcal{B}$) iff Th($\mathcal{A}$) = Th($\mathcal{B}$); this means that a first order statement is true of $\mathfrak{A}$ iff it is true in $\mathfrak{B}$.

Evidently any two isomorphic structures are elementarily equivalent. On the other hand any two dense linear orderings without endpoints or any two algebraically closed fields of characteristic zero are elementarily equivalent, although they are often nonisomorphic.

§2. n-types and saturation.

Definition 6. (Extension by constants). Let $\mathfrak{A}$ be a structure with associated first order language L. Let X be a subset or subsequence of $\mathfrak{A}$. Let $\mathfrak{A}_X$ be the structure $\mathfrak{A}$ with each element of X taken as a distinguished element. $\mathfrak{A}_X$ is essentially just $\mathfrak{A}$, but the associated language L_X is far richer than the language of $\mathfrak{A}$, as it contains names for each element of X. Thus the real field $\mathbb{R}$, as an ordered field, is customarily taken to have $\{0,1\}$ as its only distinguished elements. If on the other hand we also distinguish π and e, significantly more can be said.

Definition 7. Let $\mathfrak{A}$ be a structure, X a subset or subsequence of the domain A of $\mathfrak{A}$. Then the type of X in $\mathfrak{A}$ (in symbols: $tp_{\mathfrak{A}}(X)$, or just $tp(X)$) is defined to be the theory $Th(\mathfrak{A}_X)$.

If $a_1,\ldots,a_n \varepsilon A$ we write $tp(\bar{a})$ rather than $tp(<a_1,\ldots,a_n>)$ and call it the type (or the n-type) of $\bar{a}$ in $\mathfrak{A}$.

Remark 8. For any structure $\mathfrak{A}$ and any subset X of the domain A of $\mathfrak{A}$, tp(X) is a complete consistent theory in L_X extending the theory $Th(\mathfrak{A})$.

Definition 9. Let $\mathfrak{A}$ be a structure with associated language L. Let L_C be an extension of the language L by a set C of new constants. A theory p in L_C is called a type over $\mathfrak{A}$ in C iff

1. $Th(\mathfrak{A}) \subseteq p$
2. p is complete and consistent in L_C.

If the type p is the type of some set or sequence X in $\mathfrak{A}$, we say that p is realized (by X) in $\mathfrak{A}$.

If C is a finite set with n elements, types in C are also called n-types.

Lemma 10. With notation as in Definition 2.4, let p_o be any consistent theory in L_C such that $p_o \cup Th(\mathfrak{A})$ is consistent. Then p_o is contained in at least one n-type p.

Proof: Using Zorn's Lemma and the compactness theorem one easily extends $p_o \cup Th(\mathfrak{A})$ to a maximal consistent theory p in L_C, and it is easily seen that p is complete.

We are interested in structures $\mathfrak{A}$ which realize many types ("saturated" structures).

Definition 11. Let $\mathfrak{A}$ be a structure with language L, κ an infinite cardinal. Then $\mathfrak{A}$ is said to be κ-saturated iff for every subset X of the domain A of $\mathfrak{A}$ such that card $X < \kappa$, every 1-type p over $\mathfrak{A}_X$ is realized in $\mathfrak{A}_X$.

$\mathfrak{A}$ is saturated iff $\mathfrak{A}$ is card(A)-saturated.

Example 12. The real line $\mathbb{R}$ is an ordered set. It is not saturated, for we may consider the countable set $X = \{1/n: n=1,2,\ldots\} \subseteq \mathbb{R}$, introduce a new constant c, and consider the theory

$p_o = \{\text{"}c<1/n\text{"} : n=1,2,\ldots\} \cup \{\text{"}c>0\text{"}\}$.

It follows from the compactness theorem and Lemma 2.5 that p_o is contained in a 1-type p, and p cannot be realized in $\mathbb{R}_X$ (any realization a of p would be infinitesimal).

We will indicate a uniqueness theorem concerning saturated structures shortly. A theorem concerning the existence of sufficiently saturated structures will be deferred to §3.

Lemma 13. Let $\mathfrak{A}$ be a κ-saturated structure with language L. Let L_C be an extension of L by a set C of new constants, with card(C)$\leq\kappa$, and suppose p is a type over $\mathfrak{A}$ in C. Then p is realized in $\mathfrak{A}$.

Proof: Let $C = \{c_\alpha : \alpha<\lambda\}$ where λ is a cardinal $\leq \kappa$. Let

$C_\beta = \{c_\alpha : \alpha<\beta\}$ and let $p_\beta = p\cap L_{C_\beta}$. Then $p_o = Th(\mathcal{A})$, $p_\lambda = p$.

Claim: If p_β is realized by X_β in $\mathcal{A}$, then for some $x_\beta \varepsilon \mathcal{A}$, $p_{\beta+1}$ is realized by $X_\beta \cup \{x_\beta\}$ in $\mathcal{A}$. Evidently this claim implies the lemma, by transfinite induction.

To prove our claim, it suffices (since $\mathcal{A}$ is κ-saturated) to show that $p_{\beta+1}$ is a 1-type in $\mathcal{A}_{X_\beta}$. (If we interpret the constant c_α as the element x_α of X_β for $\alpha<\beta$, then $p_{\beta+1}$ is a theory in $(L_{C_\beta})_{\{c_\beta\}}$). The point is to check that $p_{\beta+1}$ is consistent; by the compactness theorem we may restrict our attention to a finite subset $p' = \{s_1,\ldots s_k\} \subseteq p$.

Writing more explicitly $s_i = s_i(\bar{c},c_\beta)$ with $\bar{c} \subseteq C_\beta$ define $S =$ "$\exists x\ s_1(\bar{c},x) \& S_2(\bar{c},x) \ldots \& S_k(\bar{c},x)$." Notice that $s \varepsilon p_\beta$, hence s is true in $\mathcal{A}_{X_\beta}$. Thus there is an x_β in $\mathcal{A}_{X_\beta}$ satisfying

$$s_1(\bar{x},x_\beta) \& \ldots \& \bar{s}_k(\bar{x},x_\beta).$$

Thus p' has a model, so p' is consistent, as desired.

Corollary 14. If $\mathcal{A} \equiv \mathcal{B}$ and $\mathcal{B}$ is card($\mathcal{A}$)-saturated then $\mathcal{A}$ is isomorphic with an elementary substructure of $\mathcal{B}$. (see Definition 16.)

Proof: Let $p_o = Th(\mathcal{A}_A)$. By Lemma 10 p_o is contained in a type p over B. By Lemma 13 this type is realized by a set A' of elements of B. Let $\mathcal{A}'$ be the substructure of B having domain A'. Then $\mathcal{A} \cong \mathcal{A}' < B$.

Theorem 15. (Uniqueness Theorem). Let $\mathcal{A}_1, \mathcal{A}_2$ be saturated structures of the same cardinality (i.e. card(A_1) = card (A_2)) and such that $\mathcal{A}_1 \equiv \mathcal{A}_2$. Then $\mathcal{A}_1 \cong \mathcal{A}_2$.

Proof: As the proof is a bit tedious we refer to [11,43]. The idea is to construct an isomorphism $h: \mathcal{A}_1 \cong \mathcal{A}_2$ as the union of carefully selected partial isomorphisms $h_\alpha : \mathcal{A}_1^\alpha \cong \mathcal{A}_2^\alpha$ where $\mathcal{A}_1^\alpha, \mathcal{A}_2^\alpha$ are substructures of $\mathcal{A}_1, \mathcal{A}_2$ of small cardinality. Compare the proof of Lemma 2.8 and the references cited.

§3. Unions of Chains.

Large structures can frequently be built up as directed limits of small ones. We will examine one of the more useful possibilities here, and use it in the manufacture of saturated models.

Definition 16. Let $\mathcal{A} = \langle A; \{f_i\}, \{R_j\}, \{C_k\}\rangle$, $\mathcal{B} = \langle B; \{g_i\}, \{S_j\}, \{d_k\}$ be two structures. We say that $\mathcal{A}$ is a substructure of $\mathcal{B}$ iff

1. $A \subseteq B$.
2. g_i restricted to A is f_i.
3. S_i restricted to A is R_i.
4. $d_k = c_k$.

This agrees with standard usage in our usual example, the category of ordered abelian groups.

A much stronger notion is useful in model theory. We say that $\mathcal{A}$ is an elementary substructure of $\mathcal{B}$ (in symbols: $\mathcal{A} \prec \mathcal{B}$) iff $\mathcal{A} \subseteq \mathcal{B}$ and $\mathrm{Th}(\mathcal{A}_A) = \mathrm{Th}(\mathcal{B}_A)$. As we have indicated above, $\mathcal{A}_A$ is the structure with each element distinguished (and hence named) while $\mathcal{B}_A$ is $\mathcal{B}$ with the elements of $\mathcal{A}$ distinguished.

A chain C (linearly directed system) of structures is simply an indexed set $C = \{\mathcal{A}_u : u \in U\}$ of structures $\mathcal{A}_u$, where the index set carries a linear ordering, and where $\mathcal{A}_u \subseteq \mathcal{A}_v$ for $u \leq v$. Similarly an elementary chain C is a chain $\{\mathcal{A}_u : u \in U\}$ such that $\mathcal{A}_u \prec \mathcal{A}_v$ for $u \leq v$.

Given a chain C of structures, the union $\cup C$ is defined in a thoroughly transparent fashion; if $\mathcal{A}_u = \langle A_u, \{f_i^u\}, \{R_j^u\}, \{c_k\}\rangle$, let $C = \langle \bigcup_u A_u, \{\bigcup_u f_i^u\}, \{\bigcup_u f_i^u\}, \{c_k\}\rangle$.

Theorem 17. Let C be an elementary chain. Then each structure $\mathcal{A}_u$ in C is an elementary substructure of the union $\cup C$.

Proof: Let L_u be the language of $\mathcal{A}_{u_{A_u}}$, and let $L = \bigcup_u L_u$.

Since it is evident that each $\mathfrak{A}_n$ is a <u>substructure</u> of UC, it remains to prove:

<u>Claim</u>. For any $s \in L_u$, s holds in $\mathfrak{A}_{u_{A_u}}$ iff s holds in UC. To prove this claim we use induction on the complexity of sentences. To do so it is necessary to specify rather precisely what is and what is not a first order sentence, which we have not been willing to do here. Nonetheless the argument can be sketched here:

1. First the claim is verified for the simplest sentences (e.g. "$f_i\ a_1 \ldots a_{n_i} = f_{i'} a_1' \ldots a_{n_{i'}}'$").

2. From the truth of the claim for uncomplicated sentences, we deduce its truth for more complex sentences. As is frequently the case in such arguments, the most interesting step involves the passage from sentences $s(\bar{a},b)$, to the more complex

$$\exists x\ s(\bar{a},x).$$

Assume then that our claim is valid for sentences $s(\bar{a},b)$ of some definite form; we must verify it for sentences of the form $\exists x\ s(\bar{a},x)$.

<u>Case 1</u>. In $\mathfrak{A}_u$ $\exists x\ s(\bar{a},x)$ is true. Then some specific instance $s(\bar{a},b)$ is true. By the induction hypothesis, $s(\bar{a},b)$ is true in UC, and thus $\exists x\ s(\bar{a},b)$ holds in UC.

<u>Case 2</u>. In UC $\exists x\ s(\bar{a},x)$ is true. Then $s(\bar{a},b)$ is true for some b in UC. Fix a v larger than u such that b is in $\mathfrak{A}_v$. By our induction hypothesis, $s(\bar{a},b)$ holds in $\mathfrak{A}_v$. Thus $\exists x\ s(\bar{a},v)$ holds in $\mathfrak{A}_v$. Since $\mathfrak{A}_u \prec \mathfrak{A}_v$, it follows that $\exists x\ s(\bar{a},x)$ holds also in $\mathfrak{A}_u$, as desired. ⊣

The main theorem concerning the existence of "enough" saturated structures is the following:

<u>Theorem 18</u> (Existence of saturated structures). Let κ be a regular cardinal. Let $\mathfrak{A}$ be a structure. Then there is a κ-saturated $\mathfrak{B}$ such that $\mathfrak{A} \prec \mathfrak{B}$, in brief $\mathfrak{A}$ has a κ-saturated elementary extension.

The main ingredient in the proof is the following:

Lemma 19. Let $\mathfrak{A}$ be a structure, c a constant not in the language L_A of $\mathfrak{A}_A$. Then there is an elementary extension $\mathfrak{A}'$ of $\mathfrak{A}$ such that every type over $\mathfrak{A}_A$ in $(L_A)_c$ is realized in $\mathfrak{A}'$.

Proof: For p a type over $\mathfrak{A}_A$ in $(L_A)_c$ let c_p be a new constant symbol (so that $c_p \neq c_q$ if $p \neq q$).

For any such type p let $p(c_p) = \{s(c_p): s(c) \in p\}$. Let $T = \bigcup_p p(C_p)$. We claim that T is consistent; it follows fairly easily from the compactness theorem that this is so. Then T has a model $\mathfrak{A}'$. Each element a of $\mathfrak{A}$ is named by a constant occurring in T, and that constant of course denotes an element of $\mathfrak{A}'$ as well. This induces a map $i: \mathfrak{A} \to \mathfrak{A}'$; since $Th(\mathfrak{A}_A) \subseteq T$, it follows that i identifies $\mathfrak{A}$ with an elementary substructure of $\mathfrak{A}'$. Each type p is of course realized in $\mathfrak{A}'$, by the element denoted by c_p.

Proof of Theorem 18

We use Lemma 19 and Theorem 17. Define inductively:

1. $\mathfrak{A}_0 = \mathfrak{A}$
2. $\mathfrak{A}_{\alpha+1} = \mathfrak{A}'_\alpha$ (as in Lemma 19) for $\alpha < \kappa$
3. $\mathfrak{A}_\delta = \bigcup_{\alpha<\delta} \mathfrak{A}_\alpha$ for δ a limit ordinal, $\delta \leq \kappa$.

One sees easily that $\{\mathfrak{A}_\alpha : \alpha < \kappa\}$ is an elementary chain (use Theorem 3.2 and clause 3 above). In particular $\mathfrak{A} \prec \mathfrak{A}_\kappa$. On the other hand one sees easily that $\mathfrak{A}_\kappa$ is κ-saturated.

According to Theorem 17 we can easily manufacture κ-saturated structures for κ arbitrarily large. However to obtain structures which are in fact saturated in their own cardinality requires a defter touch; we need the cardinal transfer theorems of §5 and some dubious set-theoretic axioms (e.g. GCH). Nonetheless saturated models are a versatile and desirable element of the model theorist's repertoire.

§4. The method of Diagrams.

Definition 20. Let L be a first order language with function symbols f_i, relation symbols R_j, and constant symbols c_k. By a term t we mean either a constant symbol c_k, or an expression of the form $f_i(t_1,\dots,t_{n_i})$ where $t_1,\dots,t_{n_i}$ are previously constructed terms. By an <u>atomic</u> sentence s we understand an expression of one of the following two forms:

1. $t_1 = t_2$
2. $R_j t_1 \dots t_{n_j}$

where the t_i are terms.

A <u>negatomic</u> sentence is an expression $\neg s$, where s is an atomic sentence.

If $\mathfrak{A}$ is a structure with language L, then the <u>diagram</u> of $\mathfrak{A}$ (Diag($\mathfrak{A}$)) is the following theory in L_A:

$\{s(a_1,\dots,a_n): s$ is true of $a_1,\dots,a_n$ in $\mathfrak{A}\} \cup$

$\cup\{\neg s(a_1,\dots,a_n): s$ is false of $a_1,\dots,a_n$ in $\mathfrak{A}\}$. (s atomic.)

<u>Remark 21</u>. Let $\mathfrak{A}$ be a structure, Diag ($\mathfrak{A}$) its diagram. Then a structure $\mathfrak{B}$ is a model of Diag ($\mathfrak{A}$) iff $\mathfrak{B}$ contains a substructure isomorphic to $\mathfrak{A}$.

<u>Proof</u>: If $h:\mathfrak{A} \cong \mathfrak{A}' \subseteq \mathfrak{B}$, interpret each constant a of L_A as denoting h(a) in B. Evidently B then satisfies Diag ($\mathfrak{A}$). The converse is similar.

Numerous results may be obtained by combining the Compactness Theorem with the use of diagrams. For instance:

<u>Theorem 22</u>. Let T be a theory, and let $C=\{\mathfrak{A}_n\}$ be a chain of structures such that each $\mathfrak{A}_n$ is contained in some model of T. Then $\cup C$ is a substructure of some model of T.

<u>Proof</u>. Our assumption is that $T \cup \mathrm{Diag}(\mathfrak{A}_u)$ is consistent for each u, and we claim that $T \cup \mathrm{Diag}(\cup C)$ is also consistent. Since $\mathrm{Diag}(\cup C) = \bigcup_u \mathrm{Diag}(\mathfrak{A}_u)$, this is immediate by the Compactness Theorem.

We further illustrate the method of diagrams with an application to the study of "persistent" sentences.

Definition 23. Let L be a language, T a theory in L, s a sentence of L, $\mathfrak{A}$ a model of T. Then s is said to be existential iff s has the form: $\exists x_1 \ldots x_k \, q(x_1,\ldots,x_k)$ where the formula q contains no quantifiers.

s is said to be persistent over $\mathfrak{A}$ relative to T iff s is true in every extension of $\mathfrak{A}$ which is a model of T.

Lemma 24. Retaining the notation of Definition 21, assume s is persistent over $\mathfrak{A}$ relative to T. Then there is an existential statement e in the language L which is true in $\mathfrak{A}$ such that every model of T satisfies: $e \Rightarrow s$.

Proof: $T \cup \mathrm{Diag}(\mathfrak{A}) \cup \{\neg s\}$ is inconsistent, since every model of T which extends $\mathfrak{A}$ must satisfy s. By the Compactness Theorem, there is a finite subset $F = \{s_1,\ldots s_k\} \subseteq \mathrm{Diag}(\mathfrak{A})$ such that $T \cup F \cup \{\neg s\}$ is inconsistent. Write more explicitly

$s_i = s_i(a_1,\ldots,a_r)$, exhibiting the constants

a_j of s_i which belong to $L_A - L$.

Let $e = \exists x_1 \ldots x_r \; s_1(\bar{x}) \& \ldots \& s_k(\bar{x})$. Since $T \cup F \cup \{\neg s\}$ is inconsistent, $T \cup \{e, \neg s\}$ is also inconsistent, or in other words every model of T satisfies "$e \Rightarrow s$", as desired.

§5. Cardinal Transfer Theorems.

Theorem 25. (Lowenheim-Skolem-Tarski) Let T be a theory, κ an infinite cardinality, at least as large as the cardinality of T. If T has at least one infinite model, then T has a model of cardinality κ.

Proof: It is relatively easy to see that T has a model of cardinality at least κ. Simply extend the language L of T by adjoining a set C of κ new constants, and let $T' = T \cup \{\text{"}c \neq d\text{"} : c, d \in C, c \neq d\}$.

It is easy to see that if T has an infinite model then T' has a model (use the Compactness Theorem), and evidently each model of T' has cardinality at least κ.

To complete the proof of Theorem 25 we may cite the following more precise result (which may be applied to any large model of T):

Theorem 26. Let $\mathfrak{A}$ be a structure with language L, κ an infinite cardinality such that card $(L) \leq \kappa \leq \mathrm{card}(\mathfrak{A})$. Then $\mathfrak{A}$ has an elementary substructure $\mathfrak{A}' \prec \mathfrak{A}$ of size κ.

We will not prove Theorem 26, as we have no use for the methods required. See the discussion of Skolem functions in [11,43].

Corollary 27. Let $\mathfrak{F}_1, \mathfrak{F}_2$ be two algebraically closed fields of the same characteristic p (a prime or zero). Then $\mathfrak{F}_1 \equiv \mathfrak{F}_2$.

Proof. Suppose on the contrary that s is a sentence true of $\mathfrak{F}_1$ and false of $\mathfrak{F}_2$. Let T be the theory of algebraically closed fields of characteristic p. Thus each of the theories $T \cup \{s\}, T \cup \{\neg s\}$ has an infinite model. By Theorem 5.1 we may take a model $\mathfrak{F}_1'$ of $T \cup \{s\}$ having cardinality $\aleph_1$, and a model $\mathfrak{F}_2'$ of $T \cup \{\neg s\}$ having cardinality $\aleph_1$. Then $\mathfrak{F}_1'$ and $\mathfrak{F}_2'$, being two algebraically closed fields of transcendence degree $\aleph_1$ over a common prime subfield, will be isomorphic, while s is true of $\mathfrak{F}_1'$ and false of $\mathfrak{F}_2'$, a contradiction.

§6. Definable sets.

Certain notions of algebraic geometry extend to the setting of general model theory. We will deal with this theme repeatedly in several succeeding chapters, and we confine ourselves here to a few generalities by way of background.

Let $\mathfrak{A}$ be a structure with language L and let $L_{c_1,\ldots,c_n}$ be an extension of L by constants. Let $s(c_1,\ldots,c_n)$ be a sentence of

$L_{c_1,\dots,c_n}$, and for $a_1,\dots,a_n$ in $\mathfrak{A}$ let $\mathfrak{A}_{\bar{a}}$ be $\mathfrak{A}$ with the elements $a_1,\dots a_n$ distinguished, and let

$$V(s)=\{(a_1,\dots,a_n)\in A^n: \mathfrak{A}_{\bar{a}} \text{ satisfies } s(a_1,\dots,a_n)\}.$$

V is said to be the algebraic set or relation defined by s over $\mathfrak{A}$. If $n=1$ V is said to be a definable subset of $\mathfrak{A}$.

If $\mathfrak{A}$ is replaced by $\mathfrak{A}_A$, we get a larger and in some respects more satisfactory family of definable relations. Such relations are said to be definable from parameters over $\mathfrak{A}$.

It is possible to classify definable relations in terms of the complexity of the defining formula. Some terminology:

Definition 28. A string of like quantifiers -

$$\exists y_1\dots\exists y_k$$

or $$\forall z_1\dots\forall z_\ell \quad (\text{here } k\geq 1, \ell\geq 1) \text{ -}$$

is called a quantifier block, and is more briefly indicated by writing $\exists\bar{y}$, $\forall\bar{z}$, etc.

A string of alternating quantifier blocks - e.g.

$$\exists\bar{t}\forall\bar{u}\exists\bar{v}\dots\forall\bar{z}$$

is called a quantifier prefix. A sentence which, like:

(S) "$\forall x\exists y\forall z[x>0\Rightarrow(y>x\ \&\ [y>z>x])]$"

consists of a quantifier prefix followed by a formula free of quantifiers, is said to be in prenex normal form, and is classified according to the complexity of the quantifier prefix as follows: if there are n alternations of quantifiers, the sentence is said to be an E_n sentence or an A_n sentence, according as the first block of quantifiers is existential or universal. In the example just given, S is an A_3 sentence. (If no quantifiers are present we say that S is A_0, or

E_0, or simply "quantifier-free").

In point of fact algebraic geometry over fields tends to deal only with relations defined by quantifier-free formulas, and with rather special ones at that. In model theory it seems convenient to call a definable relation a variety iff it can be defined by a quantifier-free formula. We will pursue this further in Chapter 3.

We conclude with a refinement of the notion of saturation. (Compare Definition 9 et. seq.)

Definition 29. Let $\mathfrak{A}$ be a structure with associated language L. Let L_C be an extension of L by a set C of new constants. A theory p in L_C is called a partial existential type over $\mathfrak{A}$ in C if

1. Diag $(\mathfrak{A}) \subseteq p$
2. p is consistent.

For the notions of realization and partial existential n-types, cf. Definition 9.

Let κ be a regular infinite cardinal. The structure $\mathfrak{A}$ is said to be κ-existentially saturated iff for all subsets X of $\mathfrak{A}$ of cardinality less than κ and for all expansions L_C of L by fewer than κ constants, every partial existential type over $\mathfrak{A}_X$ in C is realized in $\mathfrak{A}$.

Other basic model theoretic notions may be similarly refined. For example:

Definition 30. Two structures $\mathfrak{A}$, $\mathfrak{B}$ are elementarily equivalent at level n (in symbols: $\mathfrak{A} \equiv_n \mathfrak{B}$) iff for each E_n-sentence s, s is either true in both $\mathfrak{A}$ and $\mathfrak{B}$ or false in both $\mathfrak{A}$ and $\mathfrak{B}$.

Similarly $\mathfrak{A} \prec_n \mathfrak{B}$ iff $\mathfrak{A} \subseteq \mathfrak{B}$ and $\mathfrak{A}_A \equiv_n \mathfrak{B}_A$.

The following result will be useful in Chapter V.

Theorem 31. If $\mathfrak{A} \prec_n \mathfrak{B}$ then there is a structure $\mathfrak{C}$ such that $\mathfrak{B} \prec_{n-1} \mathfrak{C}$ and $\mathfrak{A} \prec \mathfrak{C}$.

Proof: Let L be the language of $\mathfrak{A}$ and $\mathfrak{B}$, and let L_B include names for all elements of $\mathfrak{B}$. Let

$T = Th(\mathfrak{A}) \cup Th_{n-1}(\mathfrak{B})$, where $Th_{n-1}(\mathfrak{B}) = \{s \in Th(\mathfrak{B}): s$ is an A_{n-1} sentence$\}$.

Our claim is simply that T is consistent. Supposing the contrary the compactness theorem produces a finite subset S of $Th_{n-1}(\mathfrak{B})$ such that $Th(\mathfrak{A}) \cup S$ is inconsistent. Let

$$S = \{s_i(\bar{b})\},$$

where $\bar{b}$ are constants denoting elements in $\mathfrak{B}$ but not in $\mathfrak{A}$. Then the sentence $s =$ "$\exists \bar{x} \bigwedge_i s_i(\bar{x})$" is in $Th(\mathfrak{B})$. In other words the A_n sentence $S^* =$ "$\forall \bar{x} \bigvee \neg s_i(\bar{x})$" is true in $\mathfrak{A}$, but false in $\mathfrak{B}$ (the $\bar{b}$ provide a counterexample). This violates the assumption: $\mathfrak{A} \prec_n \mathfrak{B}$.

I. Transfer Theorems in Algebra

Introduction

We will use first order logic to obtain three algebraic results: a theorem on polynomial maps in §1, **the Hilbert Nullstellensatz** and the solution by Artin and Schreier of Hilbert's Seventeenth Problem in §2.

§1. Polynomial Maps on $\mathbb{C}^n$.

Theorem 1. Let $f:\mathbb{C}^n \longrightarrow \mathbb{C}^n$ be a 1-1 polynomial map from n-dimensional complex space to itself. Then f is onto. (A polynomial map is a function of the form $f(x_1,\ldots,x_n) = (p_1(\bar{x}),\ldots,p_n(\bar{x}))$ where $p_1,\ldots,p_n$ are polynomials in n variables; we will call $\sup_i \deg(p_i)$ the degree of f).

The logical fact which makes a triviality of Theorem 1 is:

Lemma 2. Let X be a first order statement in the language of fields. Then the following are equivalent:

1. X is true in all algebraically closed fields of characteristic zero.
2. X is true in some algebraically closed field of characteristic zero.
3. For every integer n there is an algebraically closed field of characteristic $p > n$ for which X is true.

We will not dwell on the proof of Lemma 2 here (see Chapter 0 §§4-5) as we are more interested in seeing it applied.

Proof of Theorem 1:

Let K be a field, and let $X_{n,d}$ be the following assertion concerning K:

"If $f: K^n \longrightarrow K^n$ is a polynomial map of degree at most d and f is 1-1 then f is onto."

It is evident that $X_{n,d}$ can be formalized by a first order statement in the language of fields.

Our assertion is that each of the statements $X_{n,d}$ holds in $\mathbb{C}$. For any n,d it is clear that $X_{n,d}$ holds in every finite field. It follows easily that $X_{n,d}$ holds in the algebraic closure $\tilde{F}_p$ of any prime Galois field F_p, because $\tilde{F}_p$ is a direct limit of finite fields. By Lemma 2 (2.3 => 2.1) $X_{n,d}$ holds in $\mathbb{C}$, as desired. $\dashv$

Lemma 2 is a classic example of a transfer principle: all first order statements true in algebraically closed fields of large characteristic are automatically true in all algebraically closed fields of characteristic zero. The bulk of Chapter II will be devoted to a more complex (and more useful) transfer principle concerning fields with valuation.

§2. The Nullstellensatz and Hilbert's Seventeenth Problem.

We begin by recalling some basic notions of algebraic geometry.

Definition 3. Let K be a field, and let $p_1,\ldots,p_k \in K[x_1,\ldots,x_n]$ be polynomials in n variables over K.

1. $K^n = \{(a_1,\ldots,a_n): a_i \text{ in } K\}$. K^n is called (affine) n-space over K.

2. $V_K(p_1,\ldots,p_k) = \{\bar{a} \in K^n: p_1(\bar{a}) = \ldots = p_k(\bar{a}) = 0\}$. $V_K(p_1,\ldots,p_k)$ is called the variety of zeroes of $p_1,\ldots,p_k$ over K. When confusion is unlikely the subscript K is dropped.

Algebraic geometry is the study of varieties.

Our first goal in this section is the following result:

Theorem 4. (Hilbert Nullstellensatz). Let $p_1,\ldots,p_k \in \mathbb{C}[x_1,\ldots,x_n]$ be polynomials in n variables with complex coefficients. Then the following are equivalent:

1. $V_{\mathbb{C}}(p_1,\ldots,p_k) = \emptyset$
2. There arepolynomials $g_i,\ldots,g_k$ in $\mathbb{C}[x_1,\ldots,x_n]$ such that $\sum g_i p_i = 1$.

The Nullstellensatz may be neatly divided into two assertions:

Theorem 5. With the notation of Theorem 4 the following are equivalent:

1. $V_{\mathbb{C}}(p_1,\ldots,p_k) = \emptyset$.
2. For every field K extending $\mathbb{C}$ $V_K(p_1,\ldots,p_k) = \emptyset$.

Theorem 6. With the notation of Theorem 4 the following are equivalent:

1. There are polynomials $g_1,\ldots,g_k$ such that $\sum g_i p_i = 1$.
2. For every field K extending $\mathbb{C}$ $V_K(p_1,\ldots,p_k) = \emptyset$.

Since Theorem 5 is a transfer theorem (transferring information about $\mathbb{C}$ to all extensions of $\mathbb{C}$) it is reasonable to look for a model theoretic proof. Theorem 5 will follow from the slightly more general lemma following (to see this, observe that the field K referred to in 5.2 may be taken to be algebraically closed without loss of generality):

Lemma 7. Let $C \subseteq K$ be a pair of algebraically closed fields. Let $p_1,\ldots,p_k$ be polynomials over C in n variables. If $V_K(p_1,\ldots,p_k)$ is nonempty then $V_C(p_1,\ldots,p_k)$ is also nonempty.

Proof: Suppose on the contrary that:

1. $C \subseteq K$ are algebraically closed fields, $p_1,\ldots,p_k \in C[\bar{x}]$.
2. $V_C(p_1,\ldots,p_k) = \emptyset$.
3. $V_K(p_1,\ldots,p_k)$ contains a point $\bar{a} \in K^n$.

Replacing K by the algebraic closure of $C(\bar{a})$, we may assume that K has finite transcendence degree d over C. This being the case, there is a chain of algebraically closed fields:

$$C = K_0 \subseteq K_1 \subseteq \dots \subseteq K_d = K$$

each of which has transcendence degree 1 over the preceding. Let i be maximal such that $V_{K_i}(p_1,\dots,p_k) = \emptyset$. Replacing C by K_i, K by K_{i+1}, we may assume that we have in addition to 1-3 above:

4. K is the algebraic closure of a transcendental extension $C(t_0)$ of C.

We will now write down axioms which "describe" the field K. In the first place we extend the usual language of fields, which contains the symbols 0, 1, $+$, $\cdot$, by additional constants $\{c_\alpha\}$ naming all the elements of C, and one further additional constant t. We then consider the theory T consisting of the following axioms:

A1. Axioms for the theory ACF of algebraically closed fields.

A2. The addition and multiplication table for C (the "diagram" of C, Chapter 0 §4).

A3. "t is transcendental over C". More precisely we write down axioms of the following form:

"$q(t) \neq 0$"- here q is a nonzero polynomial with coefficients in C.

It is clear that K is a model of T, if we interpret "t" as t_0. More importantly:

(*) Any model of T contains a copy of K, and in particular contains a point on $V(p_1,\dots,p_k)$.

In other words <u>T proves that $V(p_1,\dots,p_k)$ is nonempty</u>. It follows that some finite subset T' of T proves that $V(p_1,\dots,p_k)$ is nonempty. But we claim that <u>any</u> finite subset T' of T is true in C (if the constant t is suitably interpreted), contradicting condition 2 above.

Thus we consider an arbitrary finite subset T' of T. We must show that the constant symbol t can be interpreted in C in such a way that the sentences in T' are true. Looking at the axioms listed in groups A1-3 above, it is evident that the axioms in groups A1-2 hold in C, and do not involve the constant t. Thus we need only find an element t' of C satisfying finitely many conditions of the form A3:

$$q_i(t) \neq 0 \quad 1 \leq i \leq r, \quad q_i \in C[x].$$

Of course there are such elements t' in C, and the proof is complete.

The rest of the Nullstellensatz (Theorem 6) is proved by a purely algebraic argument:

Proof of Theorem 6:

Let I be the ideal of the polynomial ring $C[x_1,\dots,x_n]$ generated by the polynomials $p_1,\dots,p_k$. Our claim is that the following are equivalent:

1. $1 \in I$.
2. There is no point on $V_K(p_1,\dots,p_k)$ for any extension K of C.

$\neg 2 \Rightarrow \neg 1$: Assume $C \subseteq K$ and $\bar{a}$ is a point of K^n lying on $V(p_1..p_k)$. Then every polynomial in I vanishes at $\bar{a}$, so $1 \notin I$.

$\neg 1 \Rightarrow \neg 2$: If $1 \notin I$, let M be a proper maximal ideal containing I, and let $K = C[\bar{x}]/M$. Then K is a field containing C, and we claim that $V_K(p_1,\dots,p_k)$ contains a point. Indeed let $a_i = x_i/M \in K$; then

$$p_j(\bar{a}) = p_j(\bar{x})/M = 0 \quad \text{since} \quad p_j \in I \subseteq M.$$

Thus $\bar{a}$ lies on $V_K(p_1,\dots,p_k)$.

The proof of the Nullstellensatz which we have given generalizes readily to other contexts. In the rest of this section we will apply the foregoing ideas to obtain the result of Artin and Schreier on definite functions. We begin with the definitions.

Definition 8. Let F be an ordered field, $f: F^n \longrightarrow F$ a function.

1. f is positive definite (briefly: "definite") iff for all $\bar{a} \in F^n$ $f(\bar{a}) \geq 0$.
2. $V_F(f<0) = \{\bar{a} \in F^n : f(\bar{a}) < 0\}$.

Thus a function f is definite on F iff $V_F(f < 0) = \emptyset$. We will study definite rational functions f in $F(x_1,\ldots,x_n)$.

Evidently any sum of squares of rational functions is a definite rational function (over any ordered field). Hilbert's Seventeenth Problem reads: prove the converse for definite rational functions over the real field $\mathbb{R}$.

Theorem 9 (Artin-Schreier). Let $q(x_1,\ldots,x_n)$ be a definite rational function over $\mathbb{R}$. Then there are finitely many rational functions $q_1,\ldots,q_k \in \mathbb{R}(\bar{x})$ such that: $q = \sum q_i^2$.

Like the Nullstellensatz, Theorem 9 divides conveniently into two assertions corresponding to Theorems 5,6 above:

Theorem 10. With the notation of Theorem 9 the following are equivalent:

1. q is definite over $\mathbb{R}$, i.e. $V_{\mathbb{R}}(q < 0) = \emptyset$.
2. For any ordered field F extending $\mathbb{R}$, q is definite on F, i.e. $V_F(q < 0) = \emptyset$.

Theorem 11. With the notation of Theorem 9 the following are equivalent:

1. There are finitely many rational functions $q_1,\ldots,q_k \in \mathbb{R}(\bar{x})$ such that $q = \sum q_i^2$.
2. For any ordered field F extending $\mathbb{R}$, q is definite on F.

The proof of Theorem 10 is fundamentally identical with the proof of Theorem 5, with one major sociological difference: the theory of algebraically closed fields is so well known that we felt free to exploit it in various ways in the proof (and statement) of Lemma 7. At the present juncture we need the theory of real closed fields, which we therefore pause to review.

Definition 12. Let F be an ordered field. F is said to be real closed iff F satisfies the Intermediate Value Theorem for polynomials, i.e. for any polynomial $p \in F[x]$ and elements a,b in F, if $p(a) < 0 < p(b)$ then p has a root in F lying between a and b.

Facts 13.

1. Every ordered field F has a real closure F', that is F may be embedded in a real closed extension F' which is algebraic over F. Furthermore the real closure of F is unique up to F-isomorphism.

2. If F is real closed then $F[\sqrt{-1}]$ is algebraically closed.

Proof: For details consult [26].

1. It is routine to show that:

(*) every ordered field F may be embedded in a real closed field F' algebraic over F.

The uniqueness of the real closure will be sketched in an exercise.

2. This of course is the "Fundamental Theorem of Algebra." In the first place it is rather easy to see that F satisfies:

A. Every odd degree polynomial has a root.

B. For a in F either a or $-a$ has a square root.

Using B, one sees easily that all quadratic polynomials over $F[\sqrt{-1}]$ have roots. From A,B and a bit of Galois theory one can conclude that $F[\sqrt{-1}]$ is algebraically closed [26].

We remark in passing that any field satisfying A,B has a unique ordering $<$, and is real closed with respect to this ordering. (This follows easily from the following observation: if F is an ordered field and $F[\sqrt{-1}]$ is algebraically closed, then F is real closed.)

Since any ordered field is embeddable in a real closed field, we may prove Theorem 10 in the following form:

Lemma 14. Let $R \subseteq F$ be two real closed fields and let q be a rational function defined over R. If q is definite on R then q is definite on F.

Proof: Suppose on the contrary that:

1. $R \subseteq F$ are real closed fields, $q \in R(\bar{x})$.

2. $V_R(q < 0) = \emptyset$.

3. $V_F(q < 0)$ has a point $\bar{a} \in F^n$.

Without loss of generality we may assume in addition:

4. F is the real closure of a transcendental extension $R(t_0)$ of R.

(As in the proof of Lemma 7, we may certainly assume that F has finite transcendence degree over R; the further reduction to the case of transcendence degree 1 also proceeds as in Lemma 7.)

The next step is to write down axioms describing the field F. We extend the usual language of ordered fields, which involves the symbols $0,1,+,\cdot,<$, by adjoining constants $\{r_\alpha\}$ denoting the various elements of R, as well as an additional constant t. Let T be the theory consisting of the following axioms:

A1. Axioms for the theory of real closed fields.

A2. The addition and multiplication tables of R, as well as all formulas "$r < s$" where $r,s \in R$ and $r<s$.

A3. "$q(t) \neq 0$"- here q is any nonzero polynomial over R.

A4. "$t < r$" for all r in R such that $t_0 < r$ in F, and "$r < t$" for all r in R such that $r < t_0$ in F.

Certainly F satisfies T, if we interpret t as t_0. More importantly we claim:

(*) Any model of T contains a copy of F (and hence contains a point on $V(q < 0)$).

This is not quite obvious: let F' be a model of T, and let t' be the element of F' denoted by the constant t. Certainly the ordered field $R(t')$ is contained in F', and thus F' contains the real closure of $R(t')$ (by A1). It remains to be seen that $R(t_0)$ and $R(t')$ are isomorphic as ordered fields, for then F' contains a copy of F, the real closure of $R(t_0)$. The reader should verify that the ordering on $R(t')$ is determined by the axioms A4; this is rather a key point, and uses the full force of Fact 13.2 (Fundamental Theorem of Algebra).

Granted (*), we see that q is not definite on any model of T, i.e. T proves "q is not definite". Then some finite subset T' of T proves "q is not definite". However we will see momentarily that any finite subset T' of T is true in R if the constant t is suitably interpreted, which will then contradict assumption 2 above.

If T' is a finite subset of T it is an easy matter to find a suitable interpretation t' of t in R, making all sentences in T' true. The axioms in groups A1-2 are certainly true in R. Finitely many statements of the form A3-4 can only say the following:

"t is not a root of certain polynomials $p_1,\dots,p_r$ in $R[x]$ and t lies in a certain interval (a,b)."

Evidently R contains many such elements t' (just check that every interval of R is infinite).

We must also prove Theorem 11; we confine ourselves to a sketch.

Proof of Theorem 11:

Evidently 11.1 implies 11.2. Assume therefore that q is not a sum of squares in $R(\bar{x})$. We seek an ordered field F extending R in which q is not definite. Simply take $F = R(\bar{x})$. This is a field rather than an ordered field; we will order F so that the element q is negative. It will then follow that q is not definite on F; indeed the point $\bar{x} \in F^n$ lies on $V_F(q < 0)$. How then do we order F so as to have $q < 0$?

It suffices to choose a set $P \subseteq F$ of "nonnegative" elements satisfying:

1. $a^2 \in P$ for all $a \in F$.
2. P is closed under $+,\cdot$.
3. $-q \in P$.
4. $-1 \notin P$.
5. For any a in F either a or -a is in P.

The construction of P proceeds as follows:

Let P_0 be the smallest subset of F satisfying 1-3. Using the fact that q is not a sum of squares in F, one computes that $-1 \notin P_0$. Extend P_0 by Zorn's Lemma to a maximal subset of F satisfying 1-4, and verify that 5 then holds. ⊣

§3. Notes.

Lemma 2 is the poor man's Lefschetz Principle (cf. [12], where a version of the full Lefschetz Principle is described). The application of Lemma 2 to prove Theorem 1 is mentioned in passing in [5], where it is noted that the result applies equally well to injective endomorphisms of algebraic varieties.

The Artin-Schreier theory is found in [1, 3]. The idea that Theorems 5, 10 are best viewed as logical transfer theorems was vigorously developed by Abraham Robinson and others over a period of several decades; we will see variations on this theme in Chapter II §7 and in Chapters III-V.

Exercises.

§1.

1. Disprove the following putative "transfer principle": if X is a first order statement true of all finite fields then X is true of all algebraically closed fields. Prove that this principle is valid if X is an A_2-sentence (Chapter 0 §6).

§2.

2. Let F be an ordered field contained in two real closures F', F''. Show that F' is isomorphic with F'' over F. (Sketch: one sees easily that it suffices to show that for any interval (a,b) in F and for any irreducible polynomial p over F, if

p has a root in (a,b) in F' then p has a root in (a,b) in F''. One may also assume inductively: if q is a polynomial of degree less than $\deg(p)$, then all roots of q in F' or F'' lie in F. Show: without loss of generality, p' has constant sign (+ or −) on (a,b) in F. Then apply the Intermediate Value Theorem to p.)

3. Let R be a real closed field, $R(t')$, $R(t'')$ two ordered transcendental extensions of R. Suppose that t', t'' determine the same dedekind cut of R (possibly empty or all of R). Show that the natural field isomorphism $R(t') \simeq R(t'')$ is an isomorphism of ordered fields.

II. The Ax-Kochen-Ershov Transfer Principle: (Diophantine Problems over Local Fields)

Introduction.

The most striking application of the techniques illustrated in the previous section consists of the work of Ax-Kochen-Ershov on valued fields, applied in §5 to the solution (by Ax and Kochen) of Artin's problem on zeroes of polynomials over p-adic fields $\mathbb{Q}_p$. The definition and relevant facts concerning the structure of $\mathbb{Q}_p$ will be reviewed below, primarily in §2. However the main result of Ax-Kochen-Ershov, stated explicitly in §3, does not specifically concern $\mathbb{Q}_p$; rather it is a general transfer principle analogous to the Poor Man's Lefschetz Principle (Chapter I §1). The treatment of Artin's Conjecture remains the most striking application of this principle, but we include two others: Kochen's p-adic Nullstellensatz (6), and a result concerning fields of Puiseux series, that is power series with fractional exponents (§7).

§1. Valued fields.

Notation. Whenever K is a field, $K^{\times}$ will be the multiplicative group of nonzero elements of K.

Definition 1. A valued field is a triple $\langle K, Z, \text{ord}\rangle$ where K is a field, Z is an ordered abelian group ($Z \neq 0$; Z is called the value group), and   ord: $K^X \longrightarrow Z$ is a surjective map (the <u>valuation</u>) satisfying:

1. ord(ab) = ord a + ord b.

2. ord(a+b) $\geq$ inf(ord a, ord b).

The bulk of this section is devoted to a description of some of the more natural examples of valued fields. The p-adic fields will be described in §2. As a preliminary remark we record some important trivial consequences of the axioms.

<u>Proposition 2</u>. Let <K,Z,ord> be a valued field, a,b in K.

1. ord(-a) = ord(a).

2. If ord(a) $\neq$ ord(b) then ord(a+b) = inf(ord(a), ord(b)).

We will frequently use Proposition 2   without explicit mention.

The most natural examples of valued fields are fields of analytic functions f, where ord(f) denotes the order of (the zero or pole) of f at a particular point. We will sketch the following examples: meromorphic functions in the complex plane, rational functions, and formal power series.

<u>Example 3</u>. Meromorphic functions. Let K be the field of meromorphic functions in the complex plane, i.e. the field of complex analytic functions a:$\underset{\sim}{C} \longrightarrow \underset{\sim}{C}$ whose singularities are poles. Equivalently K is the field of quotients of the ring of everywhere holomorphic functions on $\underset{\sim}{C}$.

Fix a point p in $\underset{\sim}{C}$, and define the corresponding valuation ord: $K^X \longrightarrow Z$ by

ord(a) = the order of the zero or pole of a at p.

It is immediate that <K,Z,ord> is a valued field.

We define:

$O_p = \{a \in K: \text{ord}(a) \geq 0\}$

$M_p = \{a \in K: \text{ord}(a) > 0\}$

$U_p = O_p - M_p.$

O_p is the ring of functions holomorphic at p; it is called the holomorphy ring at p, or the <u>valuation ring</u> of ord in K. M_p is easily seen to be the unique maximal ideal of $\mathcal{O}_p$. The set U_p is of lesser interest.

The residue field $\mathcal{O}_p/M_p$ is also denoted $\overline{K}$. The following diagram is easily completed:

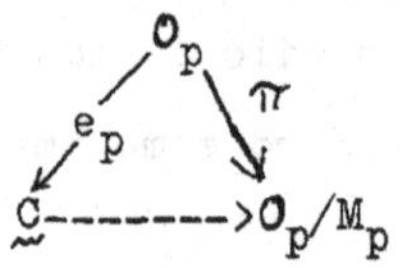

Here π is the canonical projection and e_p is the <u>evaluation map</u>: $e_p(a) = a(p)$. In other words, $\underset{\sim}{C}$ may be identified with $\overline{K}$ and the evaluation map e_p essentially coincides with the canonical projection $\mathcal{O}_p \longrightarrow \overline{K}$.

<u>Definition 4</u>. Let $\langle K, Z, \mathrm{ord}\rangle$ be a valued field. Set:

1. $\mathcal{O} = \{a \in K: \mathrm{ord}(a) \geq 0\}$. $\mathcal{O}$ is the valuation ring of ord in K.

2. $M = \{a \in \mathcal{O}: \mathrm{ord}(a) > 0\}$. M is the unique maximal ideal of $\mathcal{O}$ (check).

3. $U = \mathcal{O} - M$. This is the set of invertible elements of the ring $\mathcal{O}$. It will be of lesser interest.

4. $\overline{K} = \mathcal{O}/M$, $\pi: \mathcal{O} \longrightarrow \overline{K}$ the canonical projection.

When given a valued field K, the first order of business is to form an impression of the corresponding $\mathcal{O}$, M, $\overline{K}$.

<u>Example 5</u>. Rational functions. As a more algebraic variant of Example 3, consider the field of rational functions $K = F(x)$ over an arbitrary field F. Corresponding to any point p in F we have a valuation ord_p defined as in Example 3. Namely:

$\mathrm{ord}_p\, r(x) = n$ iff $r(x) = (x - p)^n r_1(x)$ where r_1 is a rational function defined and nonzero at p.

This of course makes sense over any field F.

The ring $\mathcal{O}_p$ and the maximal ideal M_p may be described in the manner of Example 3, and one sees that there is a diagram of the following sort:

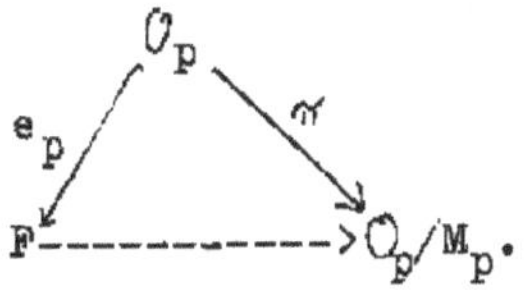

We can also develop power series expansions

$a = a_N(x-p)^N + a_{N+1}(x-p)^{N+1} + a_{N+2}(x-p)^{N+2} + \dots \quad (N = \mathrm{ord}_p\ a)$

for arbitrary rational functions a over F. The coefficients a_i may be defined inductively; they are most easily described by the condition:

(*) $a - \sum_{i=N}^{n} a_i(x-p)^i$ has order $> n$. ($N = \mathrm{ord}\ a$, the a_i are to be in F.)

Of course if F is for example the Galois field F_p with p elements, such power series expansions are purely formal.

Example 6. Formal power series. We now describe the first example of a valued field in which we will be seriously interested.

Let F be an arbitrary field and let $K = F((t))$ be the field of formal power series (Laurent series) over F. In other words the elements of $F((t))$ are formal expressions

$a = \sum_{n=N}^{\infty} a_n t^n \quad (a_n \in F,\ a_N \neq 0).$

Addition and multiplication of such expressions may be defined in the obvious way (the invertibility of nonzero power series requires checking). K is a valued field with respect to:

$\mathrm{ord}(a) = N.$

The valuation ring $\mathcal{O}$ consists of the usual formal power series (Taylor series) ring over F which may also be denoted $F[[t]]$. There is evidently a diagram of the following form:

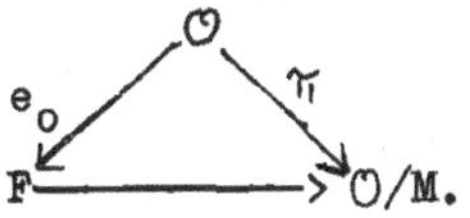

This time the evaluation map $e_o : \mathcal{O} \longrightarrow F$ is defined by $e_o(\sum_{n=0}^{\infty} a_n t^n) = a_o$.

The power series expansions described in Example 5 provide natural embeddings $i_p: F(x) \longrightarrow F((t))$. We may of course replace the formal symbol t by the formal symbol $(x-p)$, so that

$$i_p: F(x) \longrightarrow F(((x-p))).$$

The power series field $F(((x-p)))$ is the completion of $F(x)$ "at p" (see §3).

§2. p-adic Fields.

The Examples 5 and 6 of the preceding section have number theoretic analogs of great usefulness. The rational function field $F(x)$ may be replaced by the rational field $\underset{\sim}{Q}$ (or any number field; we will not descfibe the more general situation). Corresponding to the power series fields $F(("x-p"))$ we will have the p-adic fields $\underset{\sim}{Q}_p$ (described below).

Example 7. Rational numbers. Let p be a prime number. Then we may impose a valuation ord_p on the rational field Q by defining:

$ord_p(a/b) = n$ iff $a/b = p^n (a_1/b_1)$ with a_1, b_1 relatively prime to p.

The analogy with Example 5 is evident. The structure of the residue field $\mathcal{O}_p/M_p$ associated to ord_p is perhaps less evident, but it is readily seen that there is a commutative diagram:

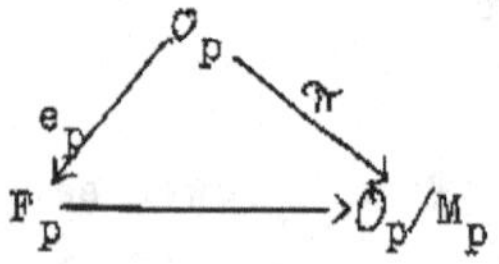

Where the map e_p is defined by $e_p(a/b) = (a \pmod p)/(b \pmod p)$.

Motivated (or carried away) by the analogy with Example 5, we may think of the rational field Q as a space of functions f with domain the set of primes and with values $f(p) = e_p(f)$ in F_p. (Exercise: graph such a "function", e.g. 3/4.) The next step is naturally to consider the "power series" expansion of a rational number a at the prime p. If for example $a \in \mathcal{O}_p$, this will be a formal expression:

(*) $a = a_0 + a_1p + a_2p^2 + a_3p^3 + \ldots \quad (0 \leq a_i < p)$

where the coefficients a_i are best described by:

(**) $a - \sum_{i=0}^{n} a_i p^i$ has order at least $n+1$.

In particular if a is a positive integer, (*) will be the usual (finite) expansion of a to the base p.

Example 8. p-adic fields. We will be particularly interested in the field of p-adic numbers $\mathbb{Q}_p$, which consists of all formal expressions

(*) $a = \sum_{n=N}^{\infty} a_n p^n \qquad (0 \leq a_n < p,\ a_N \neq 0)$.

Addition and multiplication are defined with some care. For example, in order to add two expressions $\sum a_n p^n$ and $\sum b_n p^n$ it may be necessary to reduce a_n+b_n modulo p and "carry 1" (compare the last remark in Example 7). The reader may supply a more careful definition and verify that $\mathbb{Q}_p$ is a field (it may be simpler to construct $\mathbb{Q}_p$ as the completion of $\mathbb{Q}$ "at p"- see §3).

The valuation ord on $\mathbb{Q}_p$ is defined by $\mathrm{ord}(a) = N$ (compare (*)). As usual there is a commutative diagram:

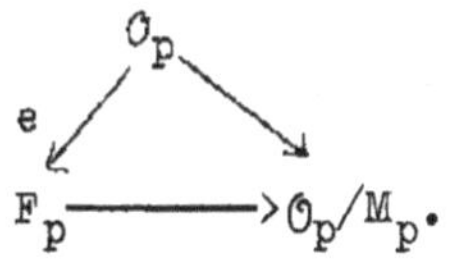

Here e is the "evaluation" map obtained by the procedure of "evaluating" a at "p = 0".

Note that at the end of Example 7 we constructed an embedding $i_p: \mathbb{Q} \longrightarrow \mathbb{Q}_p$.

We mention in passing that the fields of meromorphic functions, rational functions, or rational numbers are often called "global" fields while the power series fields and p-adic fields are called "local". The contrast in the case of fields of functions between globally defined functions and local power series expansions is evident. We note the following curious local-global transfer principle, whose model-theoretic content (if any) has never been understood.

Fact 9. (Hasse-Minkowski Principle). Let $q(\bar{x})$ be a quadratic form (homogeneous polynomial of degree 2). Let a be a rational number. Then the following are equivalent:

    1. a is in the range of q on Q.

    2. a is in the range of q on each $\underset{\sim}{Q}_p$ and on $\underset{\sim}{R}$.

This is known to be false already for cubic forms. A proof may be found in [46].

§3. Complete Fields and Hensel Fields.

Our main objective in this section is to state the Ax-Kochen-Ershov transfer principle for Hensel fields (defined below), and to verify that power series fields and p-adic fields are examples of such fields.

Definition 9. Let ⟨K,Z,ord⟩ be a valued field whose value group Z is the group of integers under addition. Let $\{a_i\}$ be a sequence of elements of K, a an element of K.

    1. a_i converges to a iff $\text{ord}(a_i-a) \longrightarrow \infty$.

    2. $\{a_i\}$ is a Cauchy sequence iff $\text{ord}(a_i-a_j) \longrightarrow \infty$ as $i,j \longrightarrow \infty$.

    3. K is complete iff every Cauchy sequence in K converges to some limit.

We make the observation that a sequence $\{a_i\}$ is Cauchy iff

    $\text{ord}(a_i-a_{i+1}) \longrightarrow \infty$ as i increases.

This follows from Definition 1, Axiom 2.

Theorem 10. Let ⟨K,Z,ord⟩ be a valued field with the group of integers as value group. Then K has an essentially unique completion K' characterized by:

    1. K' is a complete valued field.

2. K is dense in K'.

3. The functions +,· are continuous on K'.

Proof: We will merely recall the usual proof of the existence of K'. Let K' consist of all Cauchy sequences in K, identified modulo all null sequences (sequences converging to zero) and equipped with the only reasonable definitions of +,·, ord. Then K' is the completion of K, unique up to K-isomorphism. ⊣

Example 11. Any power series field $F((t))$ or p-adic field $\underset{\sim}{Q}_p$ is complete. The verification of this claim is a routine exercise, based on the circumstance that $F((t))$ and $\underset{\sim}{Q}_p$ consist by definition of all formal expressions of a certain kind (Examples 6,8).

Notice also that $F((t))$ is the completion of the rational function field $F(x)$ with respect to ord_p for any $p \in F$ (relative to the embedding $i_p: F(x) \longrightarrow F((t))$ given in Example 6) and similarly $\underset{\sim}{Q}_p$ is the completion of $\underset{\sim}{Q}$ with respect to $i_p: \underset{\sim}{Q} \longrightarrow \underset{\sim}{Q}_p$ described in Example 8. (Conditions 10.1-3 are easily verified.)

In a very definite sense there is only one nontrivial algebraic theorem about complete valued fields: Hensel's Lemma. This lemma is somewhat analogous to the Intermediate Value Theorem for polynomials over R. In the first place, both results provide roots of polynomials in complete fields; for a deeper connection see [12].

Theorem 12. (Hensel's Lemma). Let $\langle K, \underset{\sim}{Z}, \mathrm{ord} \rangle$ be a complete valued field valued in the integers. Let $p(x) \in \mathcal{O}[x]$ be a polynomial with coefficients in the valuation ring, and let $\bar{p}(x) \in \bar{K}[x]$ be the result of reducing the coefficients of p modulo the valuation ideal M. Suppose that $\bar{p}$ has a simple root α in $\bar{K}$; explicitly:

1. $\bar{p}(\alpha) = 0$.

2. $\bar{p}'(\alpha) \neq 0$.

Then p has a root a in $\mathcal{O}$ such that $\bar{a} = \alpha$.

Proof: Pick an arbitrary a_0 in $\mathcal{O}$ such that $\bar{a}_0 = \alpha$. Then the hypotheses 12.1-2 may be rephrased as follows:

1'. ord $p(a_0) > 0$.

2'. ord $p'(a_0) = 0$.

We now refine a_0 to a root of p by the method of successive approximations, relying on the completeness of K to provide the actual root. In other words we will gradually improve 1', 2' to:

I. ord $p(a) = \infty$

II. ord $p'(a) = 0$.

The basic tool here is the Taylor series expansion for polynomials.

$$(*)\ p(a_0+\varepsilon) = p(a_0) + p'(a_0)\varepsilon + \frac{p''(a_0)\varepsilon^2}{2} + \frac{p'''(a_0)\varepsilon^3}{3!} + \dots$$

$$(*')\ p'(a_0+\varepsilon) = p'(a_0) + p''(a_0)\varepsilon + \dots$$

These "Taylor series" are of course finite, since eventually all derivatives of p vanish. Thus we may view (*,*') as polynomial identities. It is now evident that if we set $\varepsilon_0 = -p(a_0)/p'(a_0)$ and $a_1 = a_0 + \varepsilon_0$, we can improve 1',2' to:

0. ord ε_0 = ord $p(a_0) > 0$.

1". ord $p(a_1) \geq 2$ ord $\varepsilon_0 = 2$ ord $p(a_0) > 0$

2". ord $p'(a_1) = 0$.

Notice that 1" really does improve 1'.

We may repeat this process, setting $\varepsilon_1 = -p(a_1)/p'(a_1)$, $a_2 = a_1+\varepsilon_1$, etc. Inspecting 1',1",... we see that

ord $p(a_n) \longrightarrow \infty$; thus

ord $\varepsilon_n \longrightarrow \infty$.

It follows that $\{a_n\}$ is a Cauchy sequence, and that the limit a of $\{a_n\}$ is a root of p. Of course $\bar{a} = \alpha$, as desired.

<u>Corollary 13</u>. Hensel's Lemma applies to the fields $F((t))$, $\mathbb{Q}_p$.

<u>Definition 14</u>. A valued field $\langle K,Z,\text{ord}\rangle$ with arbitrary value group is a <u>Hensel field</u> iff K satisfies Hensel's Lemma, i.e.:

for $p \in \mathcal{O}[x], \alpha \in \bar{K}$, if $\bar{p}(\alpha) = 0$, $\bar{p}'(\alpha) \neq 0$ then there is a root

a of p in $\mathcal{O}$ such that $\bar{a} = \alpha$.
Notice that this condition can be formalized by first order axioms in the language of valued fields.

To state the Ax-Kochen-Ershov transfer principle properly we need one moderately technical notion.

Definition 15. Let $\langle K,Z,\mathrm{ord}\rangle$ be a valued field. K is said to be unramified iff:

1. $\bar{K}$ has characteristic 0

or 2. $\bar{K}$ has characteristic $p > 0$ and ord p is the smallest positive element of Z (we express this by writing: ord p = 1).

Theorem 16 (Ax-Kochen-Ershov). Let $\langle K,Z,\mathrm{ord}\rangle$, $\langle K',Z',\mathrm{ord}'\rangle$ be two unramified Hensel fields of characteristic zero. Then the following are equivalent:

1. K is elementarily equivalent to K' (as valued fields).
2. $\bar{K}$ is elementarily equivalent to $\bar{K}'$ (as fields). Z is elementarily equivalent to Z' (as ordered groups).

Here we will assume in addition that $\bar{K}$, $\bar{K}'$ are of characteristic zero or finite.

Remark 17.

1. One defect of the above formulation of Theorem 16 is that it does not apply to p-adic completions of general number fields, where ramification occurs. The hypothesis concerning ramification can be weakened to cover this case [30].

2. Theorem 16 also applies to valued fields $K \subseteq K'$ with "elementarily equivalent" replaced everywhere by "elementarily embedded in".

An excellent exposition of the proof of Theorem 16 will be found in [30]. We will attempt a meaningful sketch of the proof. The method may be briefly described as follows (see Theorem 19):

That 16.1 implies 16.2 is a triviality. For the converse, we may assume for simplicity that K, K' (and hence $\bar{K}$, $\bar{K}'$, Z, Z') are saturated of cardinality $\aleph_1$ (Chapter 0 §2). Then $\bar{K} \simeq \bar{K}'$, $Z \simeq Z'$, and we must prove that $K \simeq K'$. In other words we must recover the structure of a saturated valued field from the structure of its residue field and value group.

To fix our ideas suppose that K is a saturated version of F((t)). The natural way to analyze K would be as follows: embed F (i.e. $\bar{K}$) in K, choose an element t of K having order 1, and form power series expansions of the elements of K. This simple idea requires a certain amount of refinement, but is essentially correct.

The construction of an isomorphism $\varphi : K \simeq K'$ proceeds in three steps. After performing the initial Step 1, the Steps 2-3 are to be iterated alternately.

<u>Step 1</u>. Embed $\bar{K} \longrightarrow U \cup \{0\}$ $(U = \mathcal{O} - M)$ as canonically as circumstances will allow. If $\bar{K}$ has characteristic 0 this is easily done via a monomorphism using Hensel's Lemma. Evidently if $\bar{K}$ has finite characteristic (while K by assumption does not) this procedure is less transparent. **(see the theory of Teichmüller representatives described in [44]).** **In any case we will embed $\bar{K}' \to U' \cup \{\bar{0}\}$ as well in such a way that the isomorphism $\bar{K}' \simeq \bar{K}'$ can be lifted to the subfields of $U \cup \{0\}$, $U' \cup \{0\}$ generated by the image of $\bar{K}$, $\bar{K}'$ respt.**

<u>Step 2</u>. Let φ_α be an isomorphism between valued subfields K_α, K'_α of K, K' with valuation group $Z_\alpha \subseteq Z$ (we have identified Z with Z' via an isomorphism). Let x be an element of $K - K_\alpha$ (or, symmetrically, of $K' - K'_\alpha$). We would like to find an isomorphism $\varphi_{\alpha+1}$ extending φ_α and having x in its domain (or range). Before dealing directly with x we will want to consider ord x in Z. Consider the group $\langle Z_\alpha, \text{ord } x\rangle \subseteq Z$. We will arrange that at each stage Z_α is a countable <u>pure</u> subgroup of Z (the latter term is defined below). In particular we should now let $Z_{\alpha+1}$ be a countable subgroup containing $\langle Z_\alpha, \text{ord } x\rangle$ and pure in Z, and then extend φ_α to an isomorphism $\varphi_{\alpha+1}$ between valued subfields $K_{\alpha+1}$, $K'_{\alpha+1}$ having value group $Z_{\alpha+1}$. Not

wishing to attempt too much at once, we have in the meantime perhaps neglected the element x. This oversight will automatically be remedied in Step 3.

Step 3. We now are considering an isomorphism $\varphi_{\alpha+1}$ between valued subfields $K_{\alpha+1}$, $K'_{\alpha+1}$ valued in a pure countable subgroup $Z_{\alpha+1}$ of Z. Consider immediate extensions of $K_{\alpha+1}$ in K; in the present context this refers to extensions K^* of $K_{\alpha+1}$ in K having the same value group $Z_{\alpha+1}$ as $K_{\alpha+1}$. We wish to extend $\varphi_{\alpha+1}$ to a maximal immediate extension of $K_{\alpha+1}$ in K. In particular it will turn out that the element x considered in Step 2 falls in the domain of our isomorphism.

The preceding sketch is immoderately vague at certain important points and we will therefore reformulate it somewhat more precisely. The following notions will be useful.

Definition 18.

1. Let $A \subseteq B$ be abelian groups. A is said to be pure in B iff for every natural number n and element a in A, if n divides a in B then n divides a already in A.

2. A crosssection χ for a valued field $\langle K,Z,\text{ord}\rangle$ is a group monomorphism $\chi : \langle Z,+\rangle \longrightarrow \langle K^{x},\cdot\rangle$ such that $\text{ord}\circ\chi = \text{identity}$. As examples we cite $\chi(n) = t^{n} \in F((t))$, $\chi(n) = p^{n} \in \mathbb{Q}_p$.

3. A crosssection χ for a valued field K is normalized iff

i. $\text{char}\,\overline{K} = 0$: no condition

ii. $\text{char}\,\overline{K} = p > 0$: $\chi(1) = p$.

4. If $K_1 \subseteq K_2$ are valued fields, K_2 is said to be an immediate extension of K_1 iff K_1, K_2 have the same value group and residue field.

We may now reformulate Theorem 16 in the form one actually intends to prove.

Theorem 19 (Proof of Theorem 16). Let $\langle K,Z,\text{ord}\rangle$, $\langle K',Z',\text{ord}'\rangle$ be two unramified saturated Hensel fields of characteristic zero and cardinality $\aleph_1$. Assume $\overline{K} \simeq \overline{K}'$, $Z \simeq Z'$. Identify Z with Z'. Then there are normalized crosssections

$\chi: Z \longrightarrow K^x$

$\chi': Z \longrightarrow K'^x$

and partial isomorphisms $\varphi_\alpha: K_\alpha \simeq K'_\alpha$ between valued subfields K_α, K'_α of K, K' ($\alpha < \aleph_1$) such that:

1. φ_β extends φ_α if $\alpha < \beta$.

2. The value group Z_α of K_α, K'_α is a countable pure subgroup of Z.

3. For z in Z, $\varphi\chi(z) = \chi'(z)$.

4. K_α has no immediate extension in K.

5. $K = \bigcup K_\alpha$, $K' = \bigcup K'_\alpha$.

In particular K,K' are isomorphic via the isomorphism $\varphi = \bigcup \varphi_\alpha$.

As we said above, we refer to [30] for the proof of Theorem 19. We do however pause in the next section to discuss cross sections further, before turning to the applications in §§5 ff.

§4. Normalized cross sections

The key to the construction of the normalized cross sections needed for Theorem 19 lies in the following result:

Theorem 27. Let A, B, U be abelian groups with U $\aleph_1$-saturated and A a pure subgroup of B. Then any homomorphism $h: A \to U$ lifts to B:

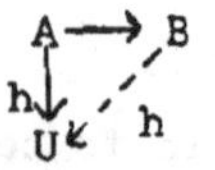

commutes for some h'.

This result is proved below (see after Lemma 29).

Corollary 28. Let <K,Z,ord> be an $\aleph_1$-saturated unramified Hensel field. Then K has a normalized cross section $\chi: \langle Z,+\rangle \to \langle K^x,\cdot\rangle$.

Proof: We chase a few diagrams (our first diagram is chased in every introduction to homological algebra).

Start with the following exact sequence of abelian groups:

$$0 \longrightarrow \langle U,\cdot\rangle \longrightarrow \langle K^x,\cdot\rangle \longrightarrow \langle Z,+\rangle \longrightarrow 0.$$

Notice that U is pure in K^x, and $\aleph_1$-saturated, so we can complete the following diagram:

$$\begin{array}{ccccc} U & \longrightarrow & K^x & \longrightarrow & Z \\ {\scriptstyle =}\downarrow & \swarrow{\scriptstyle h} & & & \\ U^{\kappa}. & & & & \end{array}$$

It follows easily that $K^x = U \oplus \ker h$, and that $\text{ord}: \ker h \cong Z$. Let χ_o be the inverse of $\text{ord}|_{\ker h}$. Then χ_o is a crosssection.

If $\text{char}\, \bar{K} = p > 0$, we normalize χ_o as follows. Consider the diagram (where $h(n\cdot 1) = p^n/\chi_o(n)$):

$$\begin{array}{ccc} \underline{Z}\cdot 1 & \longrightarrow & Z \\ h\downarrow & \swarrow{\scriptstyle h'} & \\ U^{\kappa}. & & \end{array}$$

Since $\underline{Z}\cdot 1$ is pure in Z, a suitable h' exists. Then $\chi = \chi_o \cdot h'$ is a normalized crosssection.

We turn now to the proof of Theorem 27 (see also Theorem V.20).

<u>Lemma 29</u>. If $A \subseteq B$ are abelian groups with A a pure subgroup of B and B finitely generated over A then A is a direct summand of B.

<u>Proof</u>: By the structure theorem for finite abelian groups we may write $B/A = \oplus \sum_i \langle z_i + A\rangle$ where the $z_i + A$ generate finite cyclic subgroups of B/A of order k_i. Let $y_i = k_i z_i$, so that $y_i \in A$ is divisible by k_i in B. Since A is pure in B, we can write $y_i = k_i x_i$ for some elements x_i in A. Setting $z_i' - z_i - x_i$, we have:

$$k_i z_i' = 0$$
$$z_i' + A = z_i + A.$$

It follows easily that $B = A \oplus \sum \langle z_i'\rangle$.

<u>Proof of Theorem 27</u>. Let $\{b' : b \in B\}$ be a set of constant symbols, and consider the theory

$$P = \{\textstyle\sum m_i b_i' + \sum n_j h(a_j) = 0 \ : \ b_i \in B,\ a_j \in A,\ \sum m_i b_i + \sum n_j a_j = 0\}$$

If p is consistent with the theory of U, then it is realized by elements $\{b':b\epsilon B\}\subseteq U$, and we may define $h'(b)=b'$, obtaining the desired homomorphism. To see that p is in fact consistent with Th(U), we consider an arbitrary finite subsets $p_0\subseteq p$, and show that p_0 is consistent.

Let B_0 be the subgroup of B generated over A by $\{b:$ the symbol b' occurs in $p_0\}$. Then Lemma 29 applies to $A\subseteq B_0$, so A is a direct summand of B_0. Let $\Pi:B_0\to A$ be a projection map. Then taking $h'=h\Pi$ and defining $b'=h\Pi b$ for $b\epsilon B_0$, we see that

$$\langle U;\{b':b\epsilon B_0\}\rangle$$

satisfies $Th(U)\cup p_0$, as desired.

§5. Artin's Conjecture.

The following conjecture was proposed by E. Artin.

Conjecture 33. Let K be a p-adic field $\mathbb{Q}_p$. Let $p(\bar{x})$ be a homogeneous polynomial of degree d in n variables over K. If $n > d^2$ then p has a nontrivial zero. (Of course if $\bar{x} = (0,\ldots,0)$ then $p(\bar{x}) = 0$; this is the trivial zero.)

This conjecture had been proved for power series fields $K = F_p((t))$ rather than $\mathbb{Q}_p$; it was of course understood informally that $F_p((t))$ and $\mathbb{Q}_p$ have a lot in common. Unfortunately the Artin conjecture is not always true (Terjanian's counterexample is sketched in the exercises). However the following positive result can be derived from the Ax-Kochen-Ershov principle.

Theorem 34. Fix an integer d. Then for almost all primes p, Conjecture 33 is correct for homogeneous polynomials of degree d over

$K = \mathbb{Q}_p$. The set of exceptional primes is finite and depends on d.

We break the proof into two portions, one purely algebraic and the other purely model theoretic.

Theorem 35. Conjecture 33 is correct for $K = F_p((t))$.

Theorem 36. Let S be a first order statement true of all fields $F_p((t))$. Then S is true of almost all fields $\mathbb{Q}_p$ (the exceptional set is finite and depends on S).

Theorem 34 follows directly from Theorems 35-6; simply observe that for fixed d Artin's Conjecture is formalizable by a first order statement.

Theorem 36 is a straightforward consequence of the Ax-Kochen-Ershov principle. We first note:

Theorem 37. Let T_G be a complete consistent theory of ordered abelian groups, T_F a complete theory of fields of characteristic $p \geq 0$, T the theory consisting of:

1. Axioms for the theory of valued fields $\langle K, Z, \mathrm{ord} \rangle$.
2. Axioms asserting: "Z satisfies T_G".
3. Axioms asserting: "$\bar{K}$ satisfies T_F".
4. If $p > 0$ include the axiom "ord p is the least positive element of Z".
5. If $p > 0$ we impose a side condition: T_F is the theory of a finite field F_q.
6. $\mathrm{char}(K) = 0$.

Then T is a complete theory.

Proof: It suffices to show that any two models of T are elementarily equivalent, and this is the content of the Ax-Kochen-Ershov principle (Theorem 16).

Proof of Theorem 36: Suppose toward a contradiction that S is a first order assertion true in all $F_p((t))$ and false in infinitely many fields $\mathbb{Q}_p$. Let $X = \{p: S \text{ is false in } \mathbb{Q}_p\}$.

Let T_G be the theory of $\langle \mathbb{Z}, +, < \rangle$ (the ordinary ordered group of integers). Let T_F be a complete consistent theory of fields such that:

(*) every sentence in T_F is true in infinitely many fields $\{F_p : p \in X\}$. (Such a theory T_F exists by Zorn's Lemma, and is evidently a theory of fields of characteristic zero.)

By Theorem 37 the theory T of valued fields corresponding to T_F, T_G is complete. On the other hand every finite subset of T is true in infinitely many of the fields:

$\{\underset{\sim}{Q}_p : p \in X\}$, or equally in

$\{F_p((t)) : p \in X\}$.

This follows from (*). In particular $T \cup \{S\}$, $T \cup \{\neg S\}$ are consistent, contradicting the completeness of T.

To conclude the proof of Theorem 34, we supply a purely algebraic proof of Theorem 35. The remainder of this section is devoted to this task. We will prove results concerning nontrivial roots of homogeneous polynomials: first over finite fields, then over rational function fields $F(x)$ with F finite, and finally over power series fields $F((t))$ with F finite, as desired.

Theorem 38. Let F be a finite field and let $p_1, \ldots, p_k \in F[x_1, \ldots, x_n]$ be homogeneous polynomials of degrees $d_1, \ldots, d_k$ in $n > d_1 + \ldots + d_k$ variables. Then F^n contains a nontrivial simultaneous root of $p_1, \ldots, p_k$.

Proof: Let r be the number of simultaneous roots of $p_1, \ldots, p_k$ in F^n. On account of the trivial root $r \geq 1$. Our theorem claims:

$r > 1$.

Let F have characteristic p and order q. We are going to prove that $r \equiv 0 \pmod p$, and hence $r \neq 1$, as desired. To compute r we observe that for $\bar{a}$ in F^n:

$$\prod_{i=1}^{k}(1-p_i(\bar{a})^{q-1}) = \begin{cases} 1 & p_i(\bar{a}) = 0 \text{ for } 1 \leq i \leq k \\ 0 & \text{otherwise.} \end{cases}$$

Hence r may be computed modulo p as follows:

(r) $r = \sum_{\bar{a}\in F^n} \prod_{i=1}^{k} (1-p_i(\bar{a})^{q-1}) = \sum_{F^n} \sum_{\bar{j}} c_{\bar{j}}\bar{a}^{\bar{j}} = \sum_{\bar{j}}(c_{\bar{j}} \sum_{F^n} \bar{a}^{\bar{j}})$.

Here the $c_{\bar{j}}$ are various coefficients in F and $\bar{j}$ varies over n-tuples $\bar{j} = (j_1,\dots,j_n)$; we define $\bar{a}^{\bar{j}} = \prod a_i^{j_i}$. Furthermore notice that for each $\bar{j}$:

(*) $\sum_i j_i \leq (q-1)\sum d_i < n(q-1)$.

(In other words, if $\bar{j}$ does not satisfy (*) then $c_{\bar{j}} = 0$, so the only relevant indices $\bar{j}$ are those satisfying (*).)

We complete the computation (r) above by showing that for each $\bar{j}$ satisfying (*) we have $\sum_{F^n}\bar{a}^{\bar{j}} = 0$. It suffices to note that (*) implies that for at least one i the term $j = j_i$ is less than q-1; hence:

$$\sum_{F^n} \bar{a}^{\bar{j}} = \prod_{i=1}^{n} \sum_{F} a^{j_i} = 0.$$

Here we have used the fact that $\sum_{F} a^j = 0$ for $j < q-1$, an elementary fact which is easily verified.

Theorem 39. Let K be the rational function field $F(t)$ where F is a finite field and let $p \in K[x_1,\dots,x_n]$ be a homogeneous polynomial over K of degree d in $n > d^2$ variables. Then p has a nontrivial zero in K^n.

Proof: We seek a solution of the form

$$x_i = y_{io} + y_{i1}t + \dots + y_{is}t^s$$

with $y_{ij} \in F$. We will eventually choose s rather large. We will clear the denominators from the coefficients of p, assuming therefore that p has coefficients in the polynomial ring $F[t]$. Let r be the maximum of the degrees of the coefficients of p. We may then write:

$$p(\bar{x}) = p_0(\bar{y}) + p_1(\bar{y})t + \dots + p_{ds+r}(\bar{y})t^{ds+r}.$$

We will apply Theorem 38 to get a nontrivial simultaneous zero of $p_0,\dots,p_{ds+r}$ in F, completing the proof. For this we need to have

$n(s+1) > d(ds + r + 1)$, or solving for s:

(*) $s > (d(r+1)-n)/n-d^2$.

Of course, we have just used the assumption that $n-d^2 > 0$. We may certainly choose an integer s satisfying (*); hence the proof is complete.

Proof of Theorem 35: We are given a polynomial $p(\bar{x})$ with coefficients in $F((t))$ with F a finite field. We assume that p is homogeneous of degree d in $n > d^2$ variables. We seek a nontrivial solution of $p(\bar{a}) = 0$.

We will of course reduce this problem to the previous case- $K = F(t)$. Fix a large integer k and truncate the coefficients of p above degree k, obtaining a polynomial p_k with coefficients in $F(t)$. By Theorem 39 p_k has a nontrivial root $\bar{a}_k = (a_{k1},\dots,a_{kn})$. Here $\bar{a}_k \in F(t)^n$. We would naturally like to arrange to have the roots $\bar{a}_k$ approach a nontrivial point $\bar{a}$ as $k \to \infty$.

Since the polynomials p_k are homogeneous we may multiply $\bar{a}_k$ by a suitable power of t to obtain:

(*) $\mathrm{ord}(a_{ki}) \geq 0$ for all k,i; for each k there is an i so that $\mathrm{ord}(a_{ki}) = 0$.

Now using the fact that the residue field F of $F((t))$ is finite, it is not difficult to see that it is possible to select a Cauchy subsequence of $\bar{a}_k$ (we leave the verification to the reader). Let $\bar{a}$ be the limit of such a Cauchy sequence. It follows from (*) that $\bar{a}$ is nontrivial; more precisely, for some i $\mathrm{ord}(a_i) = 0$. It is also clear that $p(\bar{a}) = 0$, as desired.

§6. Artin-Schreier Theory for p-adic Fields.

In this section we will develop a p-adic analog of the Artin-Schreier theory of real closed fields (Chapter I §2). The main result is a characterization of integral definite rational functions over p-adically closed fields (Definition 41, Theorem 45).

Theorem 40. Let T_p be the following theory of valued fields $\langle K, Z, \mathrm{ord}\rangle$:

1. Axioms for valued fields.
2. "K is a Hensel field of characteristic zero."
3. "$\overline{K} \cong F_p$" (the prime Galois field).
4. "Z satisfies the theory of $\langle \mathbb{Z}, +, <\rangle$" (the ordered group of integers).
5. "ord $p = 1$ is the smallest positive element of Z."

Then T_p is the complete theory of $\mathbb{Q}_p$ and T_p is model-complete. In particular the only nontrivial algebraic fact true about $\mathbb{Q}_p$ is Hensel's Lemma.

Proof: We apply Remark 17.3 and the definition of model-completeness (Chapter 0 §6). It suffices to prove that the theory of F_p and the theory of $\langle \mathbb{Z}, +, <\rangle$ are model complete theories. The former assertion is trivial. For the latter assertion, let T_G be the following theory of ordered abelian groups $\langle Z, +, <, 1\rangle$:

1. 1 is the smallest positive element of Z.
2. For any element z of Z and any integer $n > 0$ there is a unique "quotient" $[z/n]$ in Z and "remainder" r_n satisfying: $z = n\cdot[z/n] + r_n$, $0 \leq r_n < n\cdot 1$.

It suffices to prove that the theory T_G is model complete (the models of T_G are called Z-groups). We will indicate a suitable proof technique.

Let $Z_1 \subseteq Z_2$ be Z-groups. We will show that $Z_1 \prec Z_2$. We may assume without loss of generality that Z_1, Z_2 are saturated of cardinality $\aleph_1$. Fix any finite subset A of Z_1. We will construct an isomorphism φ of Z_1 with Z_2 fixing A:

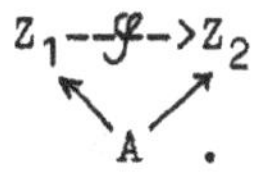

It then follows at once that $Z_1 \prec Z_2$.

Let $\langle\langle A\rangle\rangle$ be the pure subgroup of Z_1 generated by A and 1. Notice that $\langle\langle A\rangle\rangle$ is also pure in Z_2. We can in fact construct an isomorphism of Z_1 with Z_2 which fixes $\langle\langle A\rangle\rangle$. The construction of this isomorphism is left to the reader.

Definition 41.1. A valued field $\langle K,Z,\text{ord}\rangle$ is p-adically closed iff K satisfies T_p. Thus for example $\underset{\sim}{Q}_p$ is p-adically closed.

2. A rational function $r(\bar{x}) \in K(\bar{x})$ over a p-adically closed field K is integral definite iff $r: K^n \longrightarrow \mathcal{O}$. The set of all integral definite rational functions in n variables over K is denoted $I_n(K)$ (or simply I). I is a ring.

Example 42. The following rational functions are integral definite:

E1. $x/(x^2-p)$.

E2. $(x^p-x+1)^{-1}$

E3. $\gamma(x) = p^{-1}(\tau(x) - \tau(x)^{-1})^{-1}$ where $\tau(x) = x^p-x$.

Notice that the verification that these examples are in fact integral definite hinges on the following facts:

E1. $\text{ord } p = 1$ is the smallest positive element of Z.

E2. $\bar{K} = F_p$.

E3. Both of the above facts are used.

The function $\gamma(x)$ will be of fundamental importance, much as the function $\lambda(x) = x^2$ is of fundamental importance in $\underset{\sim}{R}$.

Remark 43. The ring I of integral definite functions has the following properties:

1. $\gamma \in I$.

2. If $r_1 \in I$, $r_2 \in K(\bar{x})$, then $r_1(r_2) \in I$.

3. If $r \in I$ then $(1+pr)^{-1} \in I$.

Notation 44. In conjunction with Remark 43.3 it will be convenient to introduce notation for the localization of a ring at a multiplicative set.

If R is an integral domain and T is a subset of R closed under multiplication and not containing 0, we denote by R_T the ring of quotients $\{r/t: r \in R, t \in T\}$.

The following theorem is the main result of this section:

Theorem 45. Let K be a p-adically closed field, $I = I_n(K)$ the corresponding ring of integral definite functions. Let R be the ring:

$$\mathcal{O}[\gamma[K(x_1, \ldots, x_n)]]$$

and let $T = \{1+pr: r \in R\}$. Then $I = R_T$. In other words all integral definite functions can be obtained by substitution in γ plus localization at T.

Convention. Throughout this section R, R_T are as defined above.

Notice that I certainly contains R_T. The proof that these rings are equal is somewhat devious. We prove first in the manner of Artin and Schreier that I coincides with an apparently larger ring R_T', and subsequently that $R_T' = R_T$ by a purely algebraic argument touched on in the exercises. We must first expand Remark 43 by a further significant property of I.

Definition 46. If $R \subseteq S$ are rings and $s \in S$ we say that s is integral over R iff s satisfies a monic polynomial over R. The set of elements of S integral over R forms a subring of S containing R, called the integral closure of R in S. R is integrally closed in S iff R coincides with its integral closure in S.

Lemma 47. I is integrally closed in $K(\bar{x})$.

Proof: It suffices to verify that $\mathcal{O}$ is integrally closed in K. Let a be an element of K satisfying a monic equation

$$a^n + c_{n-1}a^{n-1} + \ldots + c_o = 0$$

over $\mathcal{O}$. If $\operatorname{ord} a < 0$ it is evident that $\operatorname{ord} a^n < \operatorname{ord} c_i a^i$ for $i < n$. Hence $\operatorname{ord}(0) = \operatorname{ord} a^n < 0$, a patent contradiction. Thus $\operatorname{ord} a \geq 0$, as desired.

Definition 48. R_T' is the integral closure of R_T in $K(\bar{x})$ (notation as in Theorem 45).

Theorem 45 now becomes two theorems:

Theorem 49. $I = R_T'$.

<u>Theorem 50</u>. $R_T = R_T^!$.

We will prove Theorem 49 here. For the proof of Theorem 50 we refer to [40] (see also Exercise 13).

As a first step toward the proof of Theorem 49 it is useful to recall the proof that positive definite rational functions on $\underset{\sim}{R}$ are sums of squares (Chapter I §2). If r is not a sum of squares in $\underset{\sim}{R}(\bar{x})$, then $\underset{\sim}{R}(\bar{x})$ may be converted into an ordered field F in which r is negative. It follows easily that r is not positive definite in F, and hence not positive definite in $\underset{\sim}{R}$.

We will use the same argument in the p-adic case. In the first place we need the analog of the class of ordered fields.

<u>Definition 51</u>. $\langle K,Z,\mathrm{ord}\rangle$ is a <u>p-valued field</u> iff:

1. K is a valued field.
2. $\overline{K} = F_p$.
3. $\mathrm{ord}\ p = 1$ is the least positive element of Z.

The algebraic ingredients of the proof of Theorem 49 are as follows:

<u>Proposition 52</u>. Any p-valued field K_0 is embeddable in a p-adically closed field K.

<u>Proposition 53</u>. If K is p-adically closed, $r \in K(\bar{x})$, $r \notin R_T^!$, then $K(\bar{x})$ admits a valuation ord_0 in an ordered abelian group Z_0 in such a way that:

1. $\langle K(\bar{x}),Z_0,\mathrm{ord}_0\rangle$ is a p-valued field extending K.
2. $\mathrm{ord}_0(r) < 0$.

We prove these propositions shortly. Let us first dispose of the rest of the proof of Theorem 49.

<u>Proof of Theorem 49</u>:

We have noted that $I \supseteq R_T^!$. Assuming $r \notin R_T^!$ we will prove that r is not integral definite.

By Propositions 52-3 we may view $K(\bar{x})$ as a subfield of a p-adically closed field K^* in which $\mathrm{ord}(r) < 0$; in particular r is not integral definite on K^*. Since the theory of p-adically closed fields is model complete (Theorem 40) it follows that r is not integral definite in K, as desired.

Evidently the proof of Proposition 53 is the heart of the matter. To clear the air we will first prove Proposition 52. The following lemma is useful for the proof of Proposition 52, and provides a useful technical device in innumerable similar connections.

Lemma 54. (Place Extension Theorem). Let $\langle K, Z, ord\rangle$ be a valued field and let $K' \supseteq K$ be a field extension. Then there is a valuation ord' of K' in an ordered abelian group Z' such that $\langle K', Z', ord'\rangle$ extends $\langle K,Z,ord\rangle$.

We will prove Lemma 54 in the course of proving Proposition 53 below.

Proof of Proposition 52:

Let $\langle K_o, Z_o, ord_o\rangle$ be a p-valued field. Let K' be an algebraic closure of K_o, and apply Lemma 54 to obtain Z', ord' such that $\langle K', Z', ord'\rangle$ extends $\langle K_o, Z_o, ord_o\rangle$. Notice that Z' is a divisible group (i.e. z/n is defined for $z \in Z'$, n a natural number).

Using Zorn's Lemma choose a subgroup Z of Z' maximal with respect to:

$Z_o \subseteq Z$ and $ord\ p$ is the smallest positive element of Z.

Then let $K \subseteq K'$ be maximal subject to:

$ord'[K] \subseteq Z$, $\overline{K} = F_p$.

Let ord be ord' restricted to K.

We claim K is p-adically closed, or in more detail:

1. Z is a Z-group (defined in the proof of Theorem 40).
2. $ord[K] = Z$.
3. $\overline{K} = F_p$.
4. K is a Hensel field.

We will carry out the verification of clauses 1, 2 at some length. Of course, clause 3 was included in the definition of K.

<u>Verification 1</u>: We want to prove that for any natural number n and any z in Z, there are z', r in Z such that

(*) $z = nz' + r, \quad 0 \leq r < n\cdot 1.$

It is easily seen that we may without loss of generality assume n is prime. Furthermore our condition on r may be weakened to:

(') $r = N\cdot 1$ for some $N \geq 0$.

Thus we consider some $z \in Z$ and a prime number n. z/n exists in Z', and if $z/n \in Z$ there is nothing to prove. Otherwise it follows from the construction of Z that 1 is not the smallest positive element of $\langle Z, z/n\rangle$. In other words there is an integer m and an element z_0 of Z such that

$0 < z_0 + m\cdot z/n < 1$, or equivalently

$0 < nz_0 + mz < n\cdot 1.$

Noticing that m is relatively prime to the prime n, we fix m' so that $mm' \equiv 1 \pmod n$ and deduce:

$0 < nz_1 + z < m'n\cdot 1$ for $z_1 = m'z_0 + \frac{mm'-1}{n}\cdot 1.$

Thus $z = n(-z_1) + (z+nz_1)$, $z+nz_1 < m'n\cdot 1$,

which is the desired weak form of (*).

<u>Verification 2</u>: Let Z_1 be the <u>pure</u> subgroup of Z' generated by Z_0. Since K' is <u>algebraic</u> over K it is easy to show that $Z_1 = Z'$. This means that every element of Z'/Z_0 has finite order.

Now fix an element z of Z. We will show that z is the order of some element of K. Since z is in any case an element of Z', z has finite order over Z_0, and thus z has a finite order k over $\mathrm{ord}[K]$. In particular there is an element a of K such that $\mathrm{ord}\, a = kz$.

Let K_1 be the field $K(a^{1/k})$. We will show that $K_1 = K$, and hence $a^{1/k}$ is in K. Since $\mathrm{ord}(a^{1/k}) = z$, we will have shown that $z \in \mathrm{ord}[K]$, as desired.

To see that $K_1 = K$, it suffices to verify that:

a. $\overline{K}_1 = F_p$.

b. ord $K_1 \subseteq Z$.

Both points follow immediately from the observation that a typical element c of K_1 has the form

$$c = \sum_{i<k} c_i a^{i/k},$$

and for $i \neq j$ we have $\text{ord}\ c_i a^{i/k} \neq \text{ord}\ c_j a^{j/k}$. Thus $\text{ord}(c) \in Z$, and if $\text{ord}\ c = 0$ then $\text{ord}(c-c_0) > 0$, i.e. $\bar{c} = \bar{c}_0$.

Verification 3: By definition.

Verification 4:

We require the following general result: any valued field K has a Henselization, that is an extension K_1 which is a Hensel field, and is embeddable over K in any Hensel field containing K. Furthermore the Henselization K_1 of K is an immediate extension, i.e. K_1 has the same value group and residue field as K. In the present context we must therefore have $K=K_1$, so K is Hensel.

Henselization is discussed in a number of texts on valuation theory; we cite [59]. We also refer the reader to [60] (the machinery developed in the latter reference is essential to the proof of the Ax-Kochen-Ershov principle).

Finally we must prove Proposition 53, the key step in the proof of Theorem 49. To prove this proposition we must first understand how one goes about constructing valuations in general, and p-valuations in particular. Again we find a parallel with the theory of real closed fields, where the analogous problem is to impose an ordering on a field. Orderings of fields are conveniently constructed by choosing the set of nonnegative elements P; valuations are conveniently constructed by choosing the valuation ring $\mathcal{O}$.

<u>Definition 55</u>. Let K be a field, $\mathcal{O}$ a subring of K.

$\mathcal{O}$ is a <u>valuation ring</u> of K iff for each a K a or a^{-1} is in $\mathcal{O}$ and $\mathcal{O}$ has a unique maximal ideal $M \neq (0)$.

<u>Lemma 56</u>. Let $\mathcal{O}$ be a valuation ring of a field K with maximal ideal M. Let $U = \mathcal{O} - M$ be the multiplicative group of units in $\mathcal{O}$. Let Z be the multiplicative group K^x/U, but denote multiplication in Z by $+$. Order Z by taking $\mathcal{O}-\{0\}/U$ as the set of nonnegative elements, and let $\text{ord}: K^x \longrightarrow Z$ be the canonical projection. Then $\langle K,Z,\text{ord}\rangle$ is a valued field with valuation ring $\mathcal{O}$.

<u>Proof</u>: One verifies what needs to be verified, and no difficulties arise.

<u>Lemma 57</u>. Let A be a subring of a field K and let $I \neq 0$ be an ideal of A. Then A is contained in a valuation ring $\mathcal{O}$ of K whose unique maximal ideal M satisfies:

$I \subseteq M$.

<u>Proof</u>: By Zorn's Lemma K contains a subring $\mathcal{O}$ containing A, maximal with respect to:

(*) $I\mathcal{O} \neq \mathcal{O}$.

It suffices to show that $\mathcal{O}$ is a valuation ring of K, for then it follows easily that $I \subseteq M$.

We show first that $\mathcal{O}$ has a unique maximal ideal. Fix a maximal ideal M of $\mathcal{O}$ containing I . Let $S = \mathcal{O}-M$ and take $\mathcal{O}' = \mathcal{O}_S$ (as in Notation 44). Then $\mathcal{O} \subseteq \mathcal{O}' \subseteq K$. One verifies easily that

$I\mathcal{O}' \neq \mathcal{O}'$,

and hence $\mathcal{O} = \mathcal{O}'$. However by construction $\mathcal{O}'$ has a unique maximal ideal M', i.e. $\mathcal{O}$ has a unique maximal ideal M.

We must now verify that K is the quotient field of $\mathcal{O}$. Fix a in K. It suffices to show that either a or a^{-1} is in $\mathcal{O}$. If we assume the contrary case, it follows from the definition of $\mathcal{O}$ that

$$M\mathcal{O}[a] = \mathcal{O}[a] \quad \text{and} \quad M\mathcal{O}[a^{-1}] = \mathcal{O}[a^{-1}].$$

More explicitly we may write:

$$(+) \qquad 1 = m_0 + m_1 a + \dots + m_k a^k \qquad (m_i \in M)$$

$$(-) \qquad 1 = m_0' + m_1' a^{-1} + \dots + m_l' a^{-l} \qquad (m_i' \in M)$$

where k, l are positive integers which we may assume are minimal. Without loss of generality suppose $k \geq l$. Multiplying the first equation by $1-m_0'$ and replacing the term $(1-m_0')m_k a^k$ by

$$m_k a^k (m_1' a^{-1} + \dots + m_l' a^{-l})$$

we obtain an equation:

$$(\pm) \qquad 1-m_0' = m_0'' + m_1'' a + \dots + m_{k-1}'' a^{k-1} \quad (m_i'' \in M).$$

Since $1-m_0'$ is necessarily invertible $(\pm)$ gives rise to an equation of the form (+) with k replaced by $k-1$, contradicting the assumption that k was minimal.

This contradiction proves that K is the quotient field of $\mathcal{O}$, and concludes the proof.

In particular we are now in a position to prove Lemma 54.

Proof of Lemma 54:

We must extend a valuation from a field K to an overfield K'. Let K have valuation ring $\mathcal{O}$ and valuation ideal M. Extend $\mathcal{O}$ to a valuation ring $\mathcal{O}'$ of K' whose maximal ideal M' contains M. Then verify that the valuation constructed in Lemma 56 extends the valuation on K. The point here is that if the group of units of $\mathcal{O}'$ is denoted U', we have:

$$U' \cap \mathcal{O} = U.$$

The details are left to the reader.

With some care one can occasionally use Lemmas 56-7 to construct p-valuations, as in the following proof.

<u>Proof of Proposition 53:</u>

We have a rational function r not integral over R_T (notation as in Theorem 45). We seek a p-valuation ord on $K(\bar{x})$ such that ord $r < 0$.

Let $A = R_T[r^{-1}]$. Let $I = (p, r^{-1})$ in A. We claim first that I is a proper ideal of A. Supposing the contrary we unravel a contradiction as follows:

$-1 \in I$; explicitly:

$-1 = pa_0 + a_1 r^{-1} + a_2 r^{-2} + \ldots + a_n r^{-n}$ $\quad (a_i \in R_T)$; hence:

$0 = 1 + a_1' r^{-1} + a_2' r^{-2} + \ldots + a_n' r^{-n}$ $\quad (a_i' \in R_T)$; hence:

$r^n + a_1' r^{n-1} + \ldots + a_n' = 0$ $\quad (a_i' \in R_T)$; hence:

r is integral over R_T,

a contradiction.

Now extend A to a valuation ring $\mathcal{O}_0$ of $K(\bar{x})$ whose maximal ideal M_0 contains I. According to Lemma 56 there is a valuation ord_0 of $K(\bar{x})$ in an ordered abelian group Z_0 such that ord_0 has valuation ring $\mathcal{O}_0$. Writing K_0 for $K(\bar{x})$, we need only prove the following:

1. $\langle K_0, Z_0, \text{ord}_0\rangle$ extends $\langle K, Z, \text{ord}\rangle$.
2. $\bar{K}_0 = F_p$.
3. ord $p = 1$ is the smallest element of Z_0.

The first point is easily verified. In terms of the valuation ideals of K and K_0 we wish to prove that $M_0 \cap \mathcal{O} = M$. But by assumption $M = p\mathcal{O}$ since ord $p \equiv 1$, and p is in M_0. Thus certainly $M_0 \cap \mathcal{O} \supseteq M$, and since M is maximal $M_0 \cap \mathcal{O} = M$.

To conclude we will see that points 2,3 above follow from:

(*) $\gamma[K_0] \subseteq \mathcal{O}_0$; in other words γ is integral definite on K_0.

Verification of 2: If a is in $\mathcal{O}_0$ and $\bar{a} \notin F_p$ we may compute $\text{ord}(a^p - a) = 0$, ord $\gamma(a) \leq -1$, contradicting (*).

Verification of 3: If a is in $\mathcal{O}_0$ and $0 < \text{ord}(a) < \text{ord}\, p$, then we may compute $\operatorname{ord}(a^p - a) = \operatorname{ord} a$, $\operatorname{ord} \gamma(a) = \operatorname{ord} a - \operatorname{ord} p < 0$, contradicting (*).

§7. Puiseux Series Fields.

Definition 58. Let F be a field. A Puiseux series over F is a formal expression of the form:

(*) $a = \sum_{n \geq N} a_n t^{n/k}$ ($a_n \in F$, $a_N \neq 0$; k, N are integers and $k \geq 1$).

The set of all Puiseux series over F is denoted $F((t^{1/\infty}))$, or more briefly $P(F)$. $P(F)$ is a field with respect to the natural definitions of $+$, $\cdot$. Indeed $P(F)$ is a valued field with respect to

$\operatorname{ord}(a) = N/k$ (compare (*)).

The value group consists of the rationals under addition, and the residue field is of course F.

It is well known that if F is an algebraically closed field of characteristic zero then $P(F)$ is again algebraically closed. Oddly enough this is not the case if F is algebraically closed of finite characteristic. Again, if F is a real closed field then $P(F)$ is also a real closed field (this case can be reduced to the preceding).

In this section we prove:

Theorem 59. Let F be a p-adically closed field. Then the field of Puiseux series $P(F)$ is again a p-adically closed field.

The meaning of this theorem requires a certain amount of explanation. Call a field "weakly p-adically closed" if it is elementarily equivalent to the field $\mathbb{Q}_p$ (we say nothing for the moment about the valuation on $\mathbb{Q}_p$). Then one can prove easily that every weakly p-adically closed field K has a unique valuation ord such that K is p-adically closed with respect to ord (define the valuation ring in K as the range of Kochen's function γ). Thus the distinction between "weakly p-adically closed" fields and "p-adically closed" valued fields is inessential. (The same point arises in the theory of real closed fields; in this case the ordering $<$ can be defined in terms of the function x^2.)

Proof of Theorem 59:

A preliminary observation: by the Ax-Kochen-Ershov transfer principle $P(F)$ is elementarily equivalent to $P(\mathbb{Q}_p)$. Thus without loss of generality take $F = \mathbb{Q}_p$.

Let $\langle F^*, Z^*, ord^* \rangle$ be a proper elementary extension of $\langle F, Z, ord \rangle$. We are going to define a valuation ord' on F^* in an ordered abelian group Z' in such a way that:

1. Z' is elementarily equivalent to the ordered group of rationals.
2. The residue field $\bar{F}^*$ with respect to ord' is isomorphic with F.
3. F^* is a Hensel field with respect to ord'.

If we can do this the theorem follows at once. Indeed the Ax-Kochen-Ershov principle then says that

$\langle F^*, Z', ord' \rangle \equiv \langle P(F), Z, ord \rangle$

as valued fields, and in particular

$P(F) \equiv F^* \equiv F$

as fields.

We construct ord' as follows. Notice that $Z \subseteq Z^*$, and 1 is the smallest positive element of Z^*. Call an element a of F^* finite iff $ord\ a > n$ for some integer n (possibly negative). Let $\mathcal{O}'$ be the ring of all finite elements. $\mathcal{O}'$ has a unique maximal ideal $M' = \{a \in \mathcal{O}' : ord\ a > n$ for every integer $n\}$. The elements of M' are said to be infinitesimal. $\mathcal{O}'$ is a valuation ring of F^* (Definition 55). Let ord' be the corresponding valuation (Lemma 56). The corresponding ordered abelian group Z' consists of the quotient group $F^{*\times}/U'$ of nonzero elements of F^* modulo the units of $\mathcal{O}'$.

We must now verify that properties 1-3 above have been obtained.

Verification 1: We claim:

a. Z' is a divisible ordered abelian group, i.e. for $z \in Z'$ and $n \geq 1$ we have $z/n \in Z'$.

b. All divisible ordered abelian groups are elementarily equivalent.

This will certainly establish that Z' is elementarily equivalent to the ordered group of rationals. We leave the verification of (b) to the reader: saturate two such groups and show that they are isomorphic. To see that Z' is divisible, first check that $Z' \simeq Z^*/Z$ using the diagram:

$$\begin{array}{ccc} F^*/U^* & \longrightarrow & F^*/U' \\ \downarrow \simeq & & \downarrow \simeq \\ Z^* & \longrightarrow & Z' \end{array}$$

Since Z^* is a Z-group (as defined in the proof of Theorem 40), one checks easily that Z^*/Z is divisible.

Verification 2: The natural map (inclusion followed by canonical projection):

$$F \longrightarrow \mathcal{O}' \longrightarrow \mathcal{O}'/M'$$

provides an embedding of F into $\mathcal{O}'/M'$. We claim this map is surjective, and is thus an isomorphism. Fix an element a of $\mathcal{O}'$. We seek an element a' of F congruent to a modulo M'. Recalling that $F = \mathbb{Q}_p$, we form a power series expansion of a in $\mathbb{Q}_p$:

$$a' = \sum_{n \geq N} a_n p^n \quad (N = \text{ord}^*\, a,\ 0 \leq a_n < p)$$

with coefficients a_n characterized by:

$$\text{ord}^*(a - \sum_{n<n_0} a_n p^n) \geq n_0.$$

Then evidently a' is in F and $a-a'$ is in M', as desired.

Verification 3: To see that $\langle F^*, Z', \text{ord}' \rangle$ is a Hensel field, fix $p \in \mathcal{O}'[x]$, $a \in \mathcal{O}'$ such that

$$p(a) \in M', \quad p'(a) \notin M'.$$

We seek a root of p congruent to a modulo M'.

First consider p in the Hensel field <F*, Z*, ord*>. Then ord* p(a) is positive infinite and ord* p'(a) is not positive infinite. In particular:

(*) ord* p(a) > 2 ord* p'(a).

This allows us to use the following simple trick:
We set $\varepsilon = p(a)/p'(a)$, $q(x) = p(a + \varepsilon x)/p(a) = \sum c_j x^j$ and compute:

$c_0 = 1$, $c_1 = 1$, $\text{ord}^* c_j > 0$ for $j > 1$ (it will be seen upon inspection that the inequality (*) is just what is needed here). Thus:

$q(x) \equiv 1 + x \pmod{M^*[x]}$.

Applying Hensel's Lemma in <F*,Z*, ord*> we get a root b of q such that:

$\text{ord}^*(b+1) > 0$.

Setting $c = a + \varepsilon b$ we see that $p(c) = 0$ and $\text{ord}^*(c-a) = \text{ord}^*\varepsilon$. Thus $\text{ord}'(c-a) > 0$ and c is a root of p, proving Hensel's Lemma in <F*, Z', ord'>.

§8. Notes.

Basic references dealing with the Ax-Kochen-Ershov principle include [6,7 , 8,22,29.37] and more recently [30]

In the language of Chapter V, Theorem 27 is a special case of the following general fact: if R is a ring and M is a card(R)$^+$-saturated module over R then M is pure-injective.

The algebraic ingredients in §5 come in part from [51]. See also [32, 4]. The material in §6 was developed in [29], with the exception of Theorem 50 (the Kochen ring is integrally closed), proved in [40].

To round out the circle of ideas discussed in this chapter one should also consider pseudo-finite fields, that is fields which satisfy the theory of all finite fields. Pseudo-finite fields of characteristic zero arise in the study of the p-adic fields $\mathbb{Q}_p$ as p tends to infinity. The foundation for an excellent theory of pseudo-finite fields is provided in [5]; see also [55].

The model theory (and much of the algebra) of Hensel fields of finite characteristic consists largely of unsolved problems.

Exercises.

§1.

1. Verify Proposition 2.
2. Verify that $\mathcal{O}$ is a ring with unique maximal ideal M, and that K is the field of quotients of $\mathcal{O}$ (Definition 4).
3. If $\mathrm{ord}:K \to Z$, $\mathrm{ord}':K \to Z'$ are two valuations of a field K, we say that ord, ord' are equivalent iff the identity map on K extends to an isomorphism of $\langle K,Z,\mathrm{ord}\rangle$ with $\langle K',Z',\mathrm{ord}'\rangle$. Show that two valuations ord, ord' are equivalent iff they have the same valuation ring $\mathcal{O} = \mathcal{O}'$.
4. Find all valuations of the rational function field F(x) for which $F \subseteq \mathcal{O}$ (up to equivalence). If F is algebraically closed, is Example 5 completely general?

§2.

5. Find all valuations of the rational number field $\mathbb{Q}$ (up to equivalence).
6. Compute $\sqrt{-1}$ in $\mathbb{Q}_5$.

§3.

7. Let K be a complete valued field (Definition 9). Show that an infinite sum Σa_n exists in K iff $\lim a_n = 0$.

8. Let <K, Z, ord> be a complete valued field valued in the integers. Let p be a polynomial in $\mathcal{O}[x]$, a an element of $\mathcal{O}$ such that:

$\text{ord}\ p(a) > 2\ \text{ord}\ p'(a)$.

Prove that p has a root a' in $\mathcal{O}$.

§5.

9. (Terjanian's counterexample). Disprove Artin's conjecture. More specifically, let $p(x,y,z) =$

$x^2yz + xy^2z + xyz^2 + x^2y^2 + x^2z^2 + y^2z^2 - x^4 - y^4 - z^4$.

Define $q(x_1,\ldots,x_9) = p(x_1,x_2,x_3) + p(x_4,x_5,x_6) + p(x_7,x_8,x_9)$ and $r(x_1,\ldots,x_{18}) = q(x_1,\ldots,x_9) + 4q(x_{10},\ldots,x_{18})$.

Show:

a. If $(x,y,z) \in Z^3-(2Z)^3$ then $p(x,y,z) \equiv 3 \pmod 4$.

b. If $(x_1,\ldots,x_9) \in Z^9-(2Z)^9$ then $q(x_1,\ldots,x_9) \not\equiv 0 \pmod 4$.

c. If $\bar{x} \in \mathcal{O}_2^{18}$ and some x_i is a unit then $\text{ord}_2\, r(\bar{x}) = 0$ or 2 (and in any case, not ∞).

Thus Artin's conjecture is false for $p = 2$, $d = 4$, $n = 18$.

§6.

10. Supply the final step in the proof of Theorem 40.

11. Assume the following Theorem proved in [40]:

Let $\mathcal{O}$ be a valuation ring of $K(\bar{x})$ containing R_T with maximal ideal M, and assume $M \cap R_T = P$ is a maximal ideal of R_T; then $\mathcal{O}$ coincides with the localization of R_T at R_T-P (Notation 44). Use this to prove Theorem 50. (Hint: prove that each valuation ring of $K(\bar{x})$ is integrally closed.)

§7.

12. Generalize Theorem 59 to the case of an unramified Hensel field F valued in a Z-group.

III. Existentially Complete Structures

Introduction.

Let Σ be one of the following classes of structures: fields, ordered fields, or valued fields with residue field F_p and ord $p = 1$ (II §8). Correspondingly let $\underset{\sim}{E}_\Sigma$ be the class of algebraically closed fields, real closed fields, or p-adically closed fields. There are various ways of describing the subclass $\underset{\sim}{E}_\Sigma$ of Σ; the most natural involves the notion of existential completeness (§1). In general we may associate to any reasonable class Σ of structures a subclass $\underset{\sim}{E}_\Sigma$ of "Σ-existentially complete" structures. We simply take all structures which satisfy a suitably general analog of Hilbert's Nullstellensatz, Artin-Schreier's Theorem, or p-adic closure.

In the cases considered so far Σ is the class of models of a first order theory T, and similarly $\underset{\sim}{E}_\Sigma$ is the class of models of another first order theory T*. Life is not always so kind. In Chapter IV we will take T to be the theory of division rings, i.e. Σ = the class of all division rings, and we will see that $\underset{\sim}{E}_\Sigma$ is most definitely not the class of models of any first order theory. In particular it is somewhat difficult to give an explicit description of the existentially complete division rings.

Some terminology: when T is a first order theory and Σ is the class of all models of T, $\underset{\sim}{E}_\Sigma$ is also denoted $\underset{\sim}{E}_T$. If $\underset{\sim}{E}_T$ happens to be the class of all models of a first order theory T* we say that T* is the model companion of T and that T is companionable.

The work of Paul Cohen in set theory led Abraham Robinson to introduce two notions of "generic structure" (relative to a class Σ of structures). We speak of finitely generic and infinitely generic structures (§§2-3), and the classes of all such structures are denoted

Σ^f and Σ^∞ respectively. When Σ is the class of all models of a theory T, Σ^f and Σ^∞ are also denoted T^f and T^∞ respectively. Every finitely or infinitely generic structure is existentially complete; in other words $T^f \cup T^\infty \subseteq \mathcal{E}_T$. Whenever T is companionable we have $T^f = T^\infty = \mathcal{E}_T$. Hence these notions cast no light on the material of Chapters I-II. On the other hand the generic structures are the central objects of study in connection with uncompanionable theories.

The content of these notions- existential completeness, genericity, companionability- is best seen by considering concrete examples. In §4 we will let T be the theory of commutative rings with identity (CR), and we will see that $CR^f \cap CR^\infty = \emptyset$ (i.e. no commutative ring is both finitely and infinitely generic). In particular CR is not companionable. CR is a "bad" theory because the existentially complete models of CR have many nilpotent elements. In §5 we prove the theorem of Saracino-Lipshitz-Carson: the theory SCR of commutative rings without nilpotent elements is companionable, and in particular $SCR^f = SCR^\infty$.

The final section contains a general version of the Nullstellensatz for various classes of commutative rings.

§1. Existentially Complete Structures

Definition 1. Let Σ be a class of structures.

1. If $\mathfrak{A} \subseteq \mathfrak{B}$ are two structures we say that $\mathfrak{A}$ is existentially complete in $\mathfrak{B}$ iff for each existential sentence e defined in $\mathfrak{A}$ and true in $\mathfrak{B}$, e is true in $\mathfrak{A}$ (we allow e to contain parameters from $\mathfrak{A}$). Compare Chapter 0, §6.

2. $\mathfrak{A}$ is existentially complete (e.c.) with respect to Σ iff $\mathfrak{A}$ is in Σ and for any extension $\mathfrak{B}$ of $\mathfrak{A}$ in Σ, $\mathfrak{A}$ is existentially

complete in $\mathcal{B}$.

3. The class of existentially complete structures is denoted $\underset{\sim}{E}$ (or $\underset{\sim}{E}_\Sigma$, or- if Σ is the class of all substructures of models of a theory T- we write $\underset{\sim}{E}_T$).

We mentioned several illustrations of this concept in the Introduction to the present chapter. The following example illustrates a trivial source of pathology.

Example 2. Let Σ be the class of all ordered sets possessing first and last element (endpoints). Then $\underset{\sim}{E}_\Sigma = \emptyset$. (If a is the first element of an ordered set S, contemplate the sentence "$\exists x(x < a)$").

The following condition ensures the existence of enough existentially complete structures.

Definition 3.

1. A class Σ of structures is inductive iff the union of any chain $\{\mathcal{A}_\alpha : \mathcal{A}_\alpha \subseteq \mathcal{A}_\beta \text{ for } \alpha \leq \beta\}$ of structures from Σ is again in Σ.

2. A subclass $\Gamma \subseteq \Sigma$ is said to be cofinal with Σ iff every structure in Σ has an extension in Γ.

Theorem 4. If Σ is inductive then $\underset{\sim}{E}_\Sigma$ is cofinal with Σ.

Proof: Mimic the usual construction of an algebraically closed field or a divisible group. According to Definition 1, in order to existentially complete a given structure it may be necessary to adjoin additional elements lying in extensions. Do so. (For completeness we include some of the details as Exercise 1.)

Definition and Example 5.

If T is any first order theory let Σ_T be the class of all substructures of models of T. Then Σ_T is inductive (Chapter 0 §4 Theorem 21), so E_T contains many structures. If for example T is the theory of linearly ordered sets possessing endpoints then Σ_T consists simply of all linearly ordered sets and $\underset{\sim}{E}_T$ is easily seen to consist of all densely linearly ordered sets without endpoints. Of course in this case no structure is both existentially complete with respect to T and at the same time a model of T.

In the next two chapters we will give an adequate account of present knowledge concerning existentially complete division rings and existentially complete modules. For the present we confine ourselves to the general theory of existentially complete structures.

Our first result was used in Chapter 0 §4 to illustrate the method of diagrams (Lemma 0.4.23).

Theorem 6. Let T be a theory, $\mathfrak{A}$ a model of T, S a sentence true in all models of T extending $\mathfrak{A}$. Then for some existential sentence e true in $\mathfrak{A}$, T proves e=>S.

Proof: We briefly recall the argument. Let D be the diagram of $\mathfrak{A}$ and let T' be the theory $T \cup D \cup \{\neg S\}$. By assumption T' has no models, hence is inconsistent, hence contains a finite subset T'_0 which is inconsistent. Inspection of such a finite subset T'_0 produces the desired existential fact e true of $\mathfrak{A}$ and implying S. ⊣

Definition 7. If T is a theory, a sentence S is said to be T-persistent iff for every pair of models $\mathfrak{A} \subseteq \mathfrak{B}$ of T, if $\mathfrak{A}$ satisfies S then $\mathfrak{B}$ also satisfies S.

Example 8. Existential sentences are T-persistent for all T.

Theorem 9. Let T be a theory, S a sentence in the language of T. Then S is T-persistent iff there is an existential sentence e such that T proves S<=>e.

Proof: Let T' be the theory: $T \cup \{S\} \cup \{\neg e: T \vdash e \Rightarrow S,\ e \text{ existential}\}$. By Theorem 6 T' has no model and is therefore inconsistent. Since T' is inconsistent there must be finitely many existential sentences $e_1, \ldots, e_k$ such that:

T proves e_i=>S for each i.

$T \cup \{S\} \cup \{\neg e_1, \ldots, \neg e_k\}$ is inconsistent.

Setting $e = e_1 \vee \ldots \vee e_k$ it follows immediately that

T proves "S<=>e" and e is equivalent to an existential sentence.

<u>Lemma 10</u>. Let $\mathcal{A}$ be in Σ_T. Then $\mathcal{A}$ is existentially complete iff for every extension $\mathcal{B}$ of $\mathcal{A}$ in Σ_T there is a further extension $\mathcal{C}$ of $\mathcal{B}$ which is an elementary extension of $\mathcal{A}$. In other words the following diagram can always be completed:

$$(*)\quad \mathcal{A} \longrightarrow \mathcal{B} \dashrightarrow \mathcal{C}.$$
$$\mathcal{A} \overset{\equiv}{\dashrightarrow} \mathcal{C}$$

<u>Proof</u>: Assume first that $\mathcal{A}$ is existentially complete and that $\mathcal{B}$ in Σ_T is some extension of $\mathcal{A}$. Introduce constants naming all elements of $\mathcal{B}$ (and in particular all elements of $\mathcal{A}$) and consider the theory

$$T' = \mathrm{Th}(\mathcal{A}) \cup \mathrm{Diag}(\mathcal{B})$$

in this extended language. If T' is consistent, any model furnishes the desired $\mathcal{C}$. But T' is in fact consistent, for otherwise there would be finitely many elements $d_1, \ldots, d_k$ of $\mathrm{Diag}(\mathcal{B})$ inconsistent with $\mathrm{Th}(\mathcal{A})$. Writing $d = d_1 \,\&\, \ldots \,\&\, d_k$ and

$e = \exists x_1 \ldots \exists x_r\, d(\bar{x})$ (replace constants $\bar{b}$ from $\mathcal{B} - \mathcal{A}$ by variables $\bar{x}$)

we would then have:

e is true in $\mathcal{B}$ and false in $\mathcal{A}$,

a contradiction.

The converse is trivial. Given $\mathcal{A}$ such that $(*)$ can always be completed, a brief diagram chase shows that $\mathcal{A}$ is existentially complete.

It is occasionally useful to reformulate this lemma more precisely:

<u>Lemma 11</u>. Let $\mathcal{A} \subseteq \mathcal{B}$ be structures. Then $\mathcal{A}$ is existentially complete in $\mathcal{B}$ iff the following diagram can be completed:

$$\mathcal{A} \longrightarrow \mathcal{B} \dashrightarrow \mathcal{C}.$$
$$\mathcal{A} \overset{\equiv}{\dashrightarrow} \mathcal{C}$$

<u>Proof</u>: See the proof of Lemma 10.

<u>Definition 12</u>. Let Γ be a class of structures. A sentence S is <u>Γ-persistent</u> iff for every pair of structures $\mathcal{A} \subseteq \mathcal{B}$ in Γ, if S is

true in $\mathfrak{A}$ then S is true in $\mathfrak{B}$.

Theorem 13. Let T be a theory, S an E_3-sentence (three alternations of quantifiers in prenex normal form, with the first quantifier existential; cf. Chapter 0 §6). Then S is Σ_T-persistent.

Proof: We suppose $\mathfrak{A} \subseteq \mathfrak{B}$ are in Σ_T, and S is true in $\mathfrak{A}$. We are to conclude that S holds in $\mathfrak{B}$. Applying Lemma 10 first to $\mathfrak{A}$ and then to $\mathfrak{B}$, we may complete the following diagram:

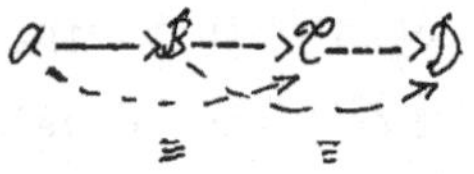

The truth of S in $\mathfrak{B}$ now follows by diagram chasing; the details are best left to the reader.

Theorem 13 is best possible (see §4).

The next two results are intended to clarify the point raised in Definition and Example 5.

Theorem 14. Let $\mathfrak{A}$ be existentially complete with respect to the theory T. Then $\mathfrak{A}$ satisfies T_{A_2} (the set of all A_2-sentences S which are consequences of T; cf. Chapter 0 §6).

Proof: Since $\mathfrak{A}$ is in Σ_T $\mathfrak{A}$ has an extension $\mathfrak{B}$ which is a model of T, and in particular of T_{A_2}. Apply Lemma 10 to complete the diagram:

For any A_2-sentence S true in $\mathfrak{B}$, it follows by diagram chasing that S is also true in $\mathfrak{A}$. In particular every sentence in T_{A_2} holds in $\mathfrak{A}$. ⊣

Theorem 15. Let T be a theory. Then T is equivalent to T_{A_2} iff the class of models of T is inductive.

Proof: If T is equivalent to T_{A_2} one shows without difficulty that the class of models of T is inductive. Assume therefore that the class of models of T is inductive. We must show that T_{A_2} proves T, or phrased more concretely: we must show that every model

$\mathfrak{A}$ of T_{A_2} is already a model of T. We do this by building a series of models:

$$(*)\quad \mathfrak{A} = \mathfrak{A}_0 \longrightarrow \mathfrak{A}_1 \longrightarrow \mathfrak{A}_2 \longrightarrow \dots \longrightarrow \mathfrak{A}_{2i} \longrightarrow \mathfrak{A}_{2i+1} \longrightarrow \mathfrak{A}_{2i+2} \longrightarrow \dots$$

with $\mathfrak{A}_{2i+1}$ a model of T for each i and such that the embeddings $\mathfrak{A}_{2i} \longrightarrow \mathfrak{A}_{2i+2}$ are elementary embeddings. If we construct (*) successfully the theorem certainly follows, for we can set $\bar{\mathfrak{A}} = \bigcup \mathfrak{A}_i$ and observe:

1. $\bar{\mathfrak{A}}$ is a model of T (since $\bar{\mathfrak{A}}$ is the union of the models $\mathfrak{A}_{2i+1}$).
2. $\bar{\mathfrak{A}}$ is an elementary extension of $\mathfrak{A}$ (since $\bar{\mathfrak{A}}$ is the union of the elementary chain $\{\mathfrak{A}_{2i}\}$; Chapter 0 §3).

Thus by 1,2 it follows that $\mathfrak{A}$ is a model of T as desired.

How does one carry out the construction of the chain (*)? Given $\mathfrak{A}_{2i}$ it suffices to construct an extension $\mathfrak{A}_{2i+1}$ of $\mathfrak{A}_{2i}$ such that

(**) $\mathfrak{A}_{2i}$ is existentially complete in $\mathfrak{A}_{2i+1}$ and $\mathfrak{A}_{2i+1}$ is a model of T.

Then Lemma 11 will supply the desired model $\mathfrak{A}_{2i+2}$ and this construction can be iterated. How then do we construct the model $\mathfrak{A}_{2i+1}$? Introducing constants naming all the elements of $\mathfrak{A}_{2i}$, we seek a model of the following theory T':

$T \cup \{u(\bar{a})$: u is any universal sentence true of the elements $\bar{a}$ of $\mathfrak{A}_{2i}\}$.

Evidently any model of this theory will satisfy (**). We therefore need only verify that T' is consistent.

Suppose on the contrary that there are finitely many universal statements $u_1(\bar{a}),\dots,u_k(\bar{a})$ true in $\mathfrak{A}_{2i}$ such that T is inconsistent with $u_1(\bar{a}) \& \dots \& u_k(\bar{a})$. Setting $u(\bar{x}) = u_1(\bar{x}) \& \dots \& u_k(\bar{x})$ we see that:

T proves $\forall \bar{x} \neg u(\bar{x})$, $\mathfrak{A}_{2i}$ satisfies $\exists \bar{x}\, u(\bar{x})$.

This contradicts the fact that $\mathcal{A}_{2i}$ satisfies T_{A_2}.

We conclude this section with two fundamental properties of $\underset{\sim}{E}_T$.

<u>Theorem 16</u>. $\underset{\sim}{E}_T$ is inductive.

<u>Proof</u>: Let $\mathcal{A}$ be the union of a chain $\{\mathcal{A}_\alpha\}$ of existentially complete structures, and let e be an existential sentence defined in $\mathcal{A}$ and true in some extension $\mathcal{B}$ of $\mathcal{A}$. Then e is already defined in some structure $\mathcal{A}_\alpha$ for reasonably large α, and is therefore true in $\mathcal{A}_\alpha$ (an existentially complete substructure of $\mathcal{B}$). Since e is persistent, e is also true in $\mathcal{A}$, as desired.

<u>Theorem 17</u>. If $\mathcal{A}$ is existentially complete in $\mathcal{B}$ and $\mathcal{B}$ is in $\underset{\sim}{E}_T$ then A is existentially complete.

<u>Proof</u>: Notice first that this theorem actually does require proof! We suppose we are given an extension $\mathcal{C}$ of $\mathcal{A}$ in which an existential sentence e is true. We want to conclude that e holds in $\mathcal{A}$. We first complete the following diagram, taking $\mathcal{D}$ to be a model of T:

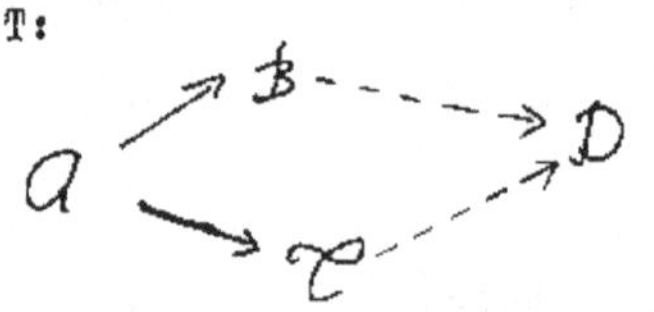

(to do this, simply verify that the theory $T \cup \text{Diag}(\mathcal{B}) \cup \text{Diag}(\mathcal{C})$ is consistent and let $\mathcal{D}$ be any model of this theory; we omit details). Then argue as follows: e holds in $\mathcal{C}$, hence in $\mathcal{D}$, hence in $\mathcal{B}$, hence in $\mathcal{A}$.

§2. <u>Infinitely Generic Structures</u>.

<u>Definition 16</u>. Let T be a theory. Then $T^E = Th(\underset{\sim}{E}_T)$ (the set of sentences true in every existentially complete structure relative to T). If $\underset{\sim}{E}_T$ is an axiomatizable class of structures, so that $\underset{\sim}{E}_T$ co-

incides with the class of models of T^E, then we say that T is <u>companionable</u>. In this case T^E is also denoted T* and is called the <u>model companion</u> of T.

Companionable theories are atypical but highly agreeable.

<u>Definition 17</u>. A class Σ of structures is <u>model complete</u> iff for any pair of structures $\mathfrak{A} \subseteq \mathfrak{B}$ in Σ, $\mathfrak{A}$ is an elementary substructure of $\mathfrak{B}$.

The following result is known as "Robinson's Test".

<u>Theorem 18</u>. If T is companionable then $\underset{\sim}{E}_T$ may be characterized as the unique subclass Γ of Σ $(= \Sigma_T)$ satisfying:

1. Γ is cofinal in Σ.
2. Γ is model complete.
3. If $\mathfrak{A} \prec \mathfrak{B} \in \Gamma$ then $\mathfrak{A} \in \Gamma$ (Γ is <u>closed</u> under "elementary substructure").

This theorem will be proved (following Corollary 31) as a consequence of more general results.

<u>Definition 19</u>. If Σ is an arbitrary class of structures and Γ is a subclass of Σ satisfying 18.1-3 then Γ is called an (in fact "the") ∞-companion of Σ.

<u>Theorem 20</u>. Let T be any theory, $\Sigma = \Sigma_T$. Then Σ has a unique ∞-companion Γ and $\Gamma \subseteq \underset{\sim}{E}_T$.

<u>Remarks</u>.

1. For most reasonably complicated theories T it is to be expected that Γ is a relatively small proper subclass of $\underset{\sim}{E}_T$, in contrast to Theorem 18. The elements of Γ are said to be <u>infinitely generic</u>.

2. Theorem 20 asserts the existence and uniqueness of the ∞-companion. As is typical in such cases the uniqueness assertion is proved by easy diagram chasing, whereas the existence assertion requires an explicit construction. There are various ways to construct the ∞-companion; none of them appears to be "canonical".

3. The proof of Theorem 20 consists of Lemma 21 through Corollary

31.

Lemma 21. Let Γ_1, Γ_2 be ∞-companions of a class Σ. Then $\Gamma_1 = \Gamma_2$.

Proof: It suffices to fix $\mathcal{A}$ in Γ_1 and prove that $\mathcal{A}$ is in Γ_2. For this purpose we build a chain:

$$\mathcal{A} = \mathcal{A}_0 \longrightarrow \mathcal{A}_1 \longrightarrow \mathcal{A}_2 \longrightarrow \ldots \longrightarrow \mathcal{A}_{2i} \longrightarrow \mathcal{A}_{2i+1} \longrightarrow \mathcal{A}_{2i+2} \longrightarrow \ldots$$

such that for each i:

$$\mathcal{A}_{2i} \in \Gamma_1, \quad \mathcal{A}_{2i+1} \in \Gamma_2.$$

This is easily carried out by induction, using property 18.1 applied to Γ_1, Γ_2.

Let $\bar{\mathcal{A}} = \bigcup \mathcal{A}_i$. By property 18.2 (model completeness of Γ_1, Γ_2) the chains $\{\mathcal{A}_{2i}\}, \{\mathcal{A}_{2i+1}\}$ are elementary chains. Hence $\mathcal{A}_0, \mathcal{A}_1$ are elementary substructures of $\bar{\mathcal{A}}$. It follows immediately that $\mathcal{A}_0$ is an elementary substructure of $\mathcal{A}_1$. By property 18.3 applied to Γ_2, $\mathcal{A} = \mathcal{A}_0 \in \Gamma_2$, as desired.

Having proved the uniqueness of the ∞-companion, we turn our attention to a construction which will establish its existence.

Definition 22.

1. Let Σ be a class of structures, $\mathcal{A} \in \Sigma$. $\mathcal{A}$ is said to be Σ-persistently complete iff for every extension $\mathcal{B}$ of $\mathcal{A}$ in Σ, every Σ-persistent sentence S true in $\mathcal{B}$ and defined in $\mathcal{A}$ is already true in $\mathcal{A}$ (compare Definition 12). The class of all Σ-persistently complete structures is denoted Σ'. By definition $\Sigma' \subseteq \Sigma$.

2. Define inductively $\Sigma^0 = \Sigma$, $\Sigma^{n+1} = (\Sigma^n)'$, $\Sigma^\infty = \bigcap \Sigma^n$.

Example 23. If $\Sigma = \Sigma_T$ then $\Sigma' = \mathcal{E}_T$ (by Theorem 9). If T is not companionable, so that $\mathcal{E}_T$ is not first order axiomatizable, then Theorem 9 does not apply to $\mathcal{E}_T$, and consequently $\Sigma'' = \mathcal{E}_T'$ is not readily identifiable.

In the present chapter we will be content to use Definition 22 as a method of obtaining the ∞-companion of Σ (Σ^∞ will be the desired ∞-companion). The real content of Definition 22 is best elucidated by algebraic examples (Chapters IV-V).

<u>Lemma 24</u>. Let Σ be an inductive class of structures. Then Σ' is inductive and cofinal with Σ.

<u>Proof</u>: Let us see first that Σ' is inductive. Let $\mathcal{A}$ be the union of a chain $\{\mathcal{A}_\alpha\}$ of Σ-persistently complete structures. Then $\mathcal{A}$ is at least in Σ. To see that $\mathcal{A}$ is again Σ-persistently complete, we consider an extension $\mathcal{B}$ of $\mathcal{A}$ in Σ satisfying a Σ-persistent sentence S defined in $\mathcal{A}$. Then for large α S is also defined in $\mathcal{A}_\alpha$, and we have:

$$\mathcal{A}_\alpha \longrightarrow \mathcal{A} \longrightarrow \mathcal{B}, \quad \mathcal{B} \text{ satisfies } S.$$

Since $\mathcal{A}_\alpha$ is in Σ', S holds in $\mathcal{A}_\alpha$. Since S is Σ-persistent, S holds in $\mathcal{A}$, as desired.

Now we must check that Σ' is cofinal with Σ. Fix $\mathcal{A}$ in Σ; we seek an extension $\mathcal{B}$ of $\mathcal{A}$ lying in Σ'. This is just an abstract version of Theorem 4. We will construct a chain:

$$\mathcal{A} = \mathcal{A}_0 \longrightarrow \mathcal{A}_1 \longrightarrow \cdots \longrightarrow \mathcal{A}_n \longrightarrow \cdots$$

with the following property:

(*) If $\mathcal{B}$ is an extension of $\mathcal{A}_{n+1}$ in Σ, and $\mathcal{B}$ satisfies some Σ-persistent sentence S defined in $\mathcal{A}_n$, then S is true in $\mathcal{A}_{n+1}$.

Then forming $\bar{\mathcal{A}} = \bigcup \mathcal{A}_n$, we see readily that $\bar{\mathcal{A}}$ is an extension of $\mathcal{A}$ in Σ'.

Our chain is constructed inductively. Given $\mathcal{A}_n$, we seek $\mathcal{A}_{n+1}$ as described in (*). $\mathcal{A}_{n+1}$ is constructed by transfinite induction from $\mathcal{A}_n$, that is we define structures $\mathcal{A}_n^\alpha$ as follows:

1. $\mathcal{A}_n^0 = \mathcal{A}_n$.
2. For limit ordinals λ, $\mathcal{A}_n^\lambda = \bigcup_{\alpha<\lambda} \mathcal{A}_n^\alpha$.
3. Given $\mathcal{A}_n^\alpha$, if $\mathcal{A}_n^\alpha$ satisfies (*) we take $\mathcal{A}_{n+1} = \mathcal{A}_n^\alpha$. Otherwise we know there is a Σ-persistent sentence S false in $\mathcal{A}_n^\alpha$ but true in an extension $\mathcal{B}$ of $\mathcal{A}_n^\alpha$. Set $\mathcal{A}_n^{\alpha+1} = \mathcal{B}$.

Evidently this construction eventually produces the desired $\mathcal{A}_{n+1}$.

<u>Corollary 25</u>. Let Σ be inductive. Then for each n Σ^n is inductive and cofinal in Σ.

<u>Proof</u>: For $n = 0$ there is nothing to prove. We argue by induction using Lemma 24.

<u>Lemma 26</u>. Let Σ be inductive. Then Σ^∞ is inductive and cofinal in Σ.

<u>Proof</u>: The inductivity of Σ^∞ follows from Corollary 25. To see that Σ^∞ is cofinal with Σ, fix $\mathfrak{A}$ in Σ and let

$$\mathfrak{A} = \mathfrak{A}_0 \longrightarrow \mathfrak{A}_1 \longrightarrow \mathfrak{A}_2 \longrightarrow \ldots \longrightarrow \mathfrak{A}_n \longrightarrow \ldots$$

be a sequence of structures such that $\mathfrak{A}_n \in \Sigma^n$ (use Corollary 25). Let $\bar{\mathfrak{A}} = \cup \mathfrak{A}_n$. Then by Corollary 25, $\bar{\mathfrak{A}} \in \Sigma^\infty$, and $\bar{\mathfrak{A}}$ extends $\mathfrak{A}$, as desired.

Thus Σ^∞ satisfies condition 18.1 from the definition of "∞-companion". We turn now to condition 18.2.

<u>Terminology 27</u>. If $u(\bar{x})$ is a formula containing variables $\bar{x}$ and $\bar{a}$ are elements of a structure $\mathfrak{A}$, we let $u(\bar{a})$ be the sentence obtained by replacing the $\bar{x}$ by (names for) the $\bar{a}$. We call $u(\bar{a})$ an <u>instance</u> of $u(\bar{x})$ (in $\mathfrak{A}$). The formula $u(\bar{x})$ is said to be Σ-persistent iff each instance of $u(\bar{x})$ is Σ-persistent.

<u>Lemma 28</u>. Let Σ be a class of structures. Then:

1. The set of Σ-persistent formulas is closed under disjunction, conjunction, and existential quantification.

2. If F is a Σ-persistent formula then $\neg F$ is a Σ'-persistent formula.

<u>Proof</u>: 28.1 is obvious and 28.2 is essentially the definition of Σ'.

<u>Lemma 29</u>. Σ^∞ is model complete.

<u>Proof</u>: Our assertion amounts to the following: every formula is Σ^∞-persistent. We may prove in fact: for every formula F there is an integer n such that F is Σ^n-persistent. Since all formulas

can be obtained from atomic formulas using conjunction, disjunction, negation, and existential quantification, it suffices to note that atomic formulas are Σ^0-persistent, and argue inductively using Lemma 28.

Having verified condition 18.2, we take up condition 18.3.

Lemma 30. Let T be a theory, Γ a cofinal subclass of Σ_T closed under elementary substructure. Then Γ' is closed under elementary substructure.

Proof: We fix an elementary substructure $\mathcal{A}$ of a structure $\mathcal{B}$ in Γ'. Since we are asked to prove that $\mathcal{A}$ is in Γ', we fix an extension $\mathcal{C}$ of $\mathcal{A}$ in which a Γ-persistent sentence S is true. We are to show that S holds already in $\mathcal{A}$. We first seek a structure $\mathcal{D}$ in Σ making the following diagram commute:

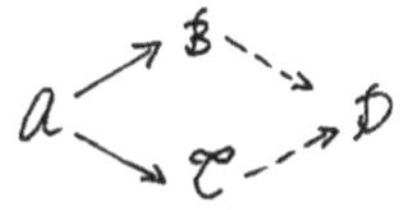

As in the proof of Theorem 17 we ask the reader to verify that $\mathcal{D}$ exists.

Since Γ is cofinal in Σ, we may assume that $\mathcal{D}$ is in Γ. Then argue as follows: S holds in $\mathcal{C}$, hence in $\mathcal{D}$ (by Γ-persistence), hence in $\mathcal{B}$ ($\mathcal{B} \in \Gamma'$), hence in $\mathcal{A}$ ($\mathcal{A} \prec \mathcal{B}$). Thus S holds in $\mathcal{A}$, as desired.

Corollary 31. If T is a theory, $\Sigma = \Sigma_T$, then Σ^∞ is closed under elementary substructure.

Proof: It suffices to prove for all n that Σ^n is closed under elementary substructure. For $n = 0$ this is clear, and Lemma 30 permits us to argue inductively.

Lemmas 21,26,29 and Corollary 31 prove Theorem 20.

Proof of Theorem 18:

Let T be companionable, i.e. $\mathbb{E}_T$ is the class of models of a theory T*. Using Theorem 9 it follows that $\mathbb{E}_T = \mathbb{E}'_T$ (check). In other words $\Sigma_T' = \Sigma_T''$, and hence by induction $\Sigma^n = \Sigma_T'$ for all $n \geq 1$. Thus $\underset{\sim}{E}_T = \Sigma_T' = \Sigma_T^\infty$, and this is the content of Theorem 18. $\dashv$

The rest of this section is devoted to a brief treatment of the connections between model completeness, completeness, and the elimination of quantifiers.

<u>Definition 32</u>. Let Σ be a class of structures.

1. Σ has the <u>joint embedding property</u> iff for any pair of structures $\mathcal{A}, \mathcal{B}$ in Σ there is a structure $\mathcal{C}$ in Σ containing both $\mathcal{A}$ and $\mathcal{B}$; in other words we can complete the diagram:

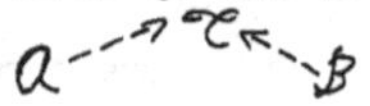

2. A structure $\mathcal{A} \in \Sigma$ is called an <u>amalgamation base</u> for Σ iff for any pair of structures $\mathcal{B}, \mathcal{C}$ in Σ extending $\mathcal{A}$ there is a structure $\mathcal{D}$ in Σ making the following diagram commute:

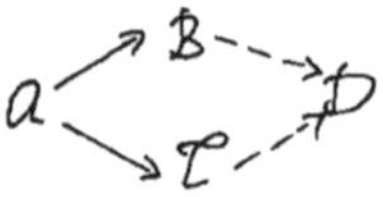

Such a structure $\mathcal{D}$ is called an <u>amalgam</u> of $\mathcal{B}$ and $\mathcal{C}$ over $\mathcal{A}$.

3. Σ has <u>amalgamation</u> iff every structure in Σ is an amalgamation base for Σ.

<u>Example 33</u>. Every existentially complete structure with respect to a theory T is an amalgamation base for Σ_T (compare the proof of Theorem 17). Typically the class of amalgamation bases is much larger than $\underset{\sim}{E}_T$.

<u>Definition 34</u>. A class of structures Σ is <u>complete</u> iff any two elements of Σ are elementarily equivalent.

<u>Theorem 35</u>. Let T be a theory, $\Sigma = \Sigma_T$. Then Σ^∞ is complete iff Σ has the joint embedding property.

<u>Proof</u>: Assume first that Σ has the joint embedding property and that $\mathcal{A}, \mathcal{B}$ are in Σ^∞. Completing the diagram:

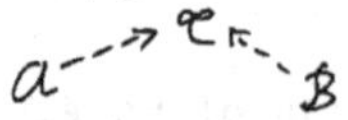

and extending $\mathcal{C}$ to a model $\mathcal{D}$ in Σ^∞, we can then use the model completeness of Σ^∞ to conclude that $\mathcal{A}, \mathcal{B}$ are elementary substructures of $\mathcal{D}$, and are thus elementarily equivalent.

Now assume Σ^∞ is complete. We are to prove that Σ has joint embedding. We fix $\mathcal{A},\mathcal{B}$ in Σ. Since Σ^∞ is cofinal with Σ we may assume that $\mathcal{A},\mathcal{B}$ are in Σ^∞ and are therefore elementarily equivalent. We now seek a joint extension of $\mathcal{A},\mathcal{B}$, that is to say we are looking for a model of

(D) $\mathrm{Diag}(\mathcal{A}') \cup \mathrm{Diag}(\mathcal{B})$.

The reader may verify that the theory (D) is consistent.

<u>Notation 36</u>. We will now have occasion to refer to <u>infinitary</u> formulas, i.e. formulas possibly involving conjunctions and disjunctions of infinitely many subformulas. We use the notation:

$\bigwedge_i F_i$, $\bigvee_j F_j$

to denote respectively the conjunction and disjunction of arbitrarily many formulas.

Infinitary formulas are not considered to be first order formulas.

<u>Theorem 37</u>. Let T be a theory, $\Sigma = \Sigma_T$. Then the following are equivalent:

1. For any first order formula $F(\bar{x})$ there are quantifier free first order formulas $q_{ij}(\bar{x})$ such that in Σ^∞ $F(\bar{x})$ is equivalent to

$\bigvee_i \bigwedge_j q_{ij}(\bar{x})$.

(This means that every structure $\mathcal{A}$ in Σ^∞ satisfies:

$\forall \bar{x}\ (F(\bar{x}) \Leftrightarrow \bigvee_i \bigwedge_j q_{ij}(\bar{x}))$.

2. Σ has amalgamation.

<u>Proof</u>: We will leave 1 => 2 as an exercise in the method of diagrams (Chapter 0 §4), analogous to the last part of the proof of Theorem 35. Assume therefore that Σ' has amalgamation and that $F(\bar{x})$ is an arbitrary first order formula. We seek an equivalent infinitary quantifier free formula.

Given $\mathcal{A}$ in Σ^∞ and elements $\bar{a}$ of $\mathcal{A}$ such that $\mathcal{A}$ satisfies

$F(\bar{a})$, let

$$q_{\mathcal{A},\bar{a}}(\bar{x}) = \bigwedge_j \{q_j(\bar{x}) : \mathcal{A} \text{ satisfies } q_j(\bar{a}), \ q \text{ quantifier free}\}.$$

Let $q(\bar{x}) = \bigvee \{q_{\mathcal{A},\bar{a}}(\bar{x}) : \mathcal{A}$ satisfies $F(\bar{a})\}$.

Then $q(\bar{x})$ is an infinitary formula of the desired form. Fixing $\mathcal{A}$ in Σ^{∞}, $\bar{a}$ in $\mathcal{A}$, we now need only verify:

(*) $\mathcal{A}$ satisfies $F(\bar{a})$ iff $\mathcal{A}$ satisfies $q(\bar{a})$.

Now clearly if $\mathcal{A}$ satisfies $F(\bar{a})$ then $\mathcal{A}$ satisfies $q_{\mathcal{A},\bar{a}}(\bar{a})$, and in particular $\mathcal{A}$ satisfies $q(\bar{a})$. For the converse assume $\mathcal{A}$ satisfies $q(\bar{a})$, so that for some $\mathcal{B}$, $\bar{b}$ we have:

1. $\mathcal{B}$ satisfies $F(\bar{b})$
2. $\mathcal{A}$ satisfies $q_{\mathcal{B},\bar{b}}(\bar{a})$.

By virtue of the second point there is an isomorphism $h : \langle\bar{a}\rangle \simeq \langle\bar{b}\rangle$. hence we may apply amalgamation to complete the diagram:

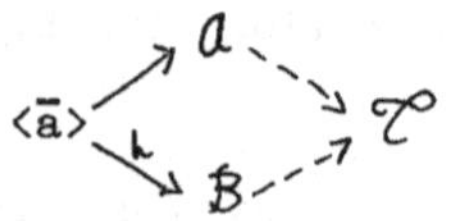

and then extend $\mathcal{C}$ to a model $\mathcal{D}$ in Σ^{∞}. Since Σ^{∞} is model complete we conclude:

$\mathcal{B}$ satisfies $F(h\bar{a})$ =>

$\mathcal{D}$ satisfies $F(h\bar{a})$ =>

$\mathcal{A}$ satisfies $F(\bar{a})$

as desired.

<u>Corollary 38</u>. If T is companionable and Σ_T has amalgamation then the model companion T^* of T has "elimination of quantifiers", i.e. for each first order formula $F(\bar{x})$ there is a quantifier free first order formula $q(\bar{x})$ such that T^* proves:

$\forall\bar{x}\ (F(\bar{x}) \Leftrightarrow q(\bar{x}))$.

<u>Proof</u>: Applying Theorems 18 and 37 we know that there is an infinitary quantifier free formula $q(\bar{x}) = \bigvee_i \bigwedge_j q_{ij}(\bar{x})$ such that T^* proves:

(*) $\forall\bar{x}\ (F(\bar{x}) \equiv q(x))$

in the sense that (*) holds in all models of T^*.

Let T_i be the theory $\{q_{ij}: j \text{ varies}\}$ (the variables $\bar{x}$ in q_{ij} may be replaced by constant symbols, so the formulas $q_{ij}(\bar{x})$ may be viewed as sentences q_{ij}). Then (*) implies in particular:

$T \cup T_i$ proves $F(\bar{x})$.

It follows that for some finite subset T_i' of T_i,

$T \cup T_i'$ proves $F(\bar{x})$.

Define:

$$q_i'(\bar{x}) = \bigwedge_j \{q_{ij}(\bar{x}): q_{ij} \in T_i'\} .$$

Then $q_i'(\bar{x})$ is a first order quantifier free formula, and T^* proves:

(**) $\forall \bar{x}\ (F(\bar{x}) \equiv \bigvee_i q_i'(\bar{x}))$.

Having replaced the infinite conjunctions in (*) by finite conjunctions in (**), we must now repeat the argument just used to reduce the infinite disjunction in (**) to a finite disjunction (details will be left to the reader). Rereading (**), and taking the disjunction $\bigvee_i$ to be finite, we see that Corollary 38 has been proved.

§3. Finitely Generic Structures.

Finitely generic structures and finite forcing will be of interest from two points of view. In this section we deal with the model theoretic aspects of finite forcing, illustrating this material in §4 with an algebraic example. **The more concrete applications of the method of finite forcing have a rather different flavor. This point is illustrated in Chapter IV §4.**

Definition 39. Let Σ be a class of structures.

1. A condition is a finite subset of the diagram of a structure in Σ, that is to say a finite set of atomic sentences $A(\bar{a})$ and negated atomic sentences $\neg B(\bar{a})$ which are true of certain elements $\bar{a}$ in some structure $\mathfrak{A}$ in Σ.

2. Let C be a condition, S a sentence. We define the relation "C <u>forces</u> S" by induction on the complexity of S:

a. If S is atomic then "C forces S" means that S is in C.

b. If $S = S_1 \vee S_2$ then "C forces S" means that C forces S_1 or C forces S_2.

c. We treat conjunctions $S_1 \& S_2$ similarly.

d. If $S = \exists x\, S_1(x)$ then "C forces S" means that for some element a occurring in a formula of C, C forces $S_1(a)$.

e. If $S = \neg S_1$ then "C forces S" means that no condition extending C forces S_1.

f. If $S = \forall x\, S_1(x)$ then "C forces S" means that C forces $\neg \exists x \neg S_1(x)$.

Example 40. Let Σ be the class of all structures equipped with a function f (and no other functions, relations, or constants). Let S be the sentence:

$$\forall y \exists x_1 \exists x_2 \, (fx_1 = y \,\&\, fx_2 = y \,\&\, x_1 \neq x_2).$$

Then S is forced by the empty condition $\emptyset$. (It is worthwhile to verify this fact by mechanically unraveling the definition of forcing.)

Definition 41. Let Σ be a class of structures, $\mathfrak{A}$ an element of Σ.

1. A sentence S is forced by $\mathfrak{A}$ iff S is forced by a condition true in $\mathfrak{A}$.

2. $\mathfrak{A}$ is finitely Σ-generic iff for every sentence S defined in $\mathfrak{A}$, S holds in $\mathfrak{A}$ iff $\mathfrak{A}$ forces S.

3. The class of all finitely Σ-generic structures is denoted Σ^f and is called the finite companion of Σ.

The reason Σ^f merits our attention at this point is that Σ^f is a subclass of $\mathbb{E}_\Sigma$, sometimes equal to Σ^∞ but not infrequently actually disjoint from Σ^∞. When $\Sigma^f \cap \Sigma^\infty = \emptyset$ the study of the differences between finitely Σ-generic and infinitely Σ-generic structures gives us a deeper understanding of $\mathbb{E}_\Sigma$. It may also be said that structures in Σ^∞ are obtained by throwing in everything but the kitchen sink (see §2) whereas the finitely generic structures are constructed with a bit more finesse.

We saw in example 5 that the entire class $\underset{\sim}{E}_{\Sigma}$ may easily be empty; hence we restricted our attention to inductive classes Σ. A second major source of pathology arises from the fact that the language of Σ may well be uncountable; there are numerous perfectly unobjectionable theories T in uncountable languages for which the finite companion Σ_T^f is empty. For various kinds of pathology and minor relations between $\underset{\sim}{E}_{\Sigma}$, Σ^f, and Σ^∞ see the exercises.

The theory of finitely generic structures is at its simplest when $\Sigma = \Sigma_T$ and the language of T is countable. We will prove our theorems in varying degrees of generality. When the language of T is uncountable the class Σ_T^f is not necessarily uninteresting; it is merely necessary to bear in mind that it may be empty.

We begin by placing some trivialities on record.

Lemma 42. Let C be a condition, S a sentence.

1. C cannot force both S and $\neg$S.
2. C has an extension C' which forces S or $\neg$S.
3. If C is contained in a condition C' and C forces S then C' also forces S.
4. If C involves certain constant symbols $\bar{a}$ and if C' is the condition obtained by replacing the constants $\bar{a}$ by new constants $\bar{b}$ not found in C or S then:

   C forces $S(\bar{a})$ iff C' forces $S(\bar{b})$.

Proof: 1,2 are immediate. 3,4 are proved mechanically by induction on the complexity of S.

Theorem 43. Let T be a theory in a countable language, C a condition relative to Σ_T. Then C is true of the elements of some finitely Σ_T-generic structure $\mathcal{A}$ in Σ_T. In particular $\Sigma_T^f \neq \emptyset$.

Proof: The construction of $\mathcal{A}$ is as follows. Fix an infinite set of constants A. Call a sentence S acceptable iff all constants occurring in S are either in A or in the language of T. Construct an increasing sequence of acceptable conditions:

$$C = C_0 \subseteq C_1 \subseteq C_2 \subseteq \ldots$$

such that for any acceptable sentence S there is an n such that

C_n forces S or $\neg S$.

(This is easily done using 42.2 and the fact that there are only countably many acceptable sentences.)

Let $D = \bigcup C_n$. We claim that D is the diagram of a finitely generic structure $\mathcal{A}$ in Σ_T.

Notice first that for any acceptable atomic sentence S, either S or $\neg S$ is in D (see the Definition 39). Thus D is the diagram of some structure $\mathcal{A}$. It is easily seen that $\mathcal{A}$ is in Σ_T, i.e. $\mathcal{A}$ is a substructure of some model of $T-$ indeed one must simply check that $D \cup T$ is consistent, and this is obvious (Chapter 0 §4).

By definition and the above construction, for every acceptable sentence S:

$\mathcal{A}$ forces either S or $\neg S$.

From this fact alone we may deduce that $\mathcal{A}$ is finitely Σ_T-generic. We prove by induction on the complexity of acceptable sentences S:

(*) $\mathcal{A}$ satisfies S iff $\mathcal{A}$ forces S.

The argument is thoroughly trivial unless S has the form $\neg S_1$, where of course we assume (*) for S_1. Even here one direction is trivial: if $\mathcal{A}$ forces S then $\mathcal{A}$ does not force S_1, hence does not satisfy S_1, and thus does satisfy S. For the converse, assume $\mathcal{A}$ satisfies S. We know that $\mathcal{A}$ forces S_1 or $\neg S_1$, but by (*) applied to S_1 $\mathcal{A}$ does not force S_1. Hence $\mathcal{A}$ forces $\neg S_1$, as desired.

Proposition 44. Let Σ be a class of structures, $\mathcal{A}$ an element of Σ. Then $\mathcal{A}$ is finitely Σ-generic iff for every sentence S defined in $\mathcal{A}$, $\mathcal{A}$ forces S or $\neg S$.

Proof: "only if" is trivial and "if" was proved above.

Theorem 45. Let Σ be a class of structures, $\mathcal{A}$ finitely Σ-generic. Then $\mathcal{A}$ is existentially complete with respect to Σ.

Proof: Suppose $\mathcal{A}$ has an extension $\mathcal{B}$ in Σ satisfying an existential sentence $e = \exists \bar{x} q(\bar{x},\bar{a})$ defined in $\mathcal{A}$ (here q is quantifier free). We claim that $\mathcal{A}$ satisfies e. Supposing the contrary, we see that $\mathcal{A}$ satisfies $\neg e$, and hence forces $\neg e$. Thus there is a condition C in $\mathcal{A}$ forcing

$\neg \exists \bar{x}\ q(\bar{x},\bar{a})$.

On the other hand we have elements $\bar{b}$ of $\mathcal{B}$ satisfying $q(\bar{b},\bar{a})$. Letting C' be the union of C with all atomic formulas or negated atomic formulas which are true of $\bar{b},\bar{a}$ and occur in q, we see that C' is true in $\mathcal{B}$, hence is a condition. On the other hand clearly C' forces e (Definition 39). This contradicts Lemma 42 since $C \subseteq C'$ and C forces $\neg e$.

Corollary 46. Let T be an inductive theory. Then every finitely T-generic model satisfies T.

Proof: In fact every Σ_T-existentially complete model satisfies T, by Theorems 14-15.

The next order of business is to see to what extent the characterization of Σ^{∞} given in Theorem 20 applies also to Σ^{f}.

Theorem 47. Σ^{f} is model complete and closed under elementary substructure. If Σ is inductive, Σ^{f} is inductive.

Proof: (This theorem is of course consistent with the possibility that $\Sigma^{f} = \emptyset$.)

To see that Σ^{f} is model complete, let $\mathcal{A} \subseteq \mathcal{B}$ be finitely Σ-generic. Then any sentence S true of $\mathcal{A}$ is forced by $\mathcal{A}$, hence clearly also by $\mathcal{B}$, and is hence true of $\mathcal{B}$. Thus $\mathcal{A}$ is an elementary substructure of $\mathcal{B}$, as desired.

Now suppose that $\mathcal{A}$ is an elementary substructure of a structure $\mathcal{B}$ in Σ^{f}. We claim that $\mathcal{A}$ is in Σ^{f}. We apply Proposition 45. Fix a sentence $S(\bar{a})$ defined in $\mathcal{A}$; we will verify that S or $\neg S$

is forced by $\mathcal{A}$. Since $\mathcal{B}$ is finitely Σ-generic there is a condition $C(\bar{a},\bar{b})$ true in $\mathcal{B}$ and forcing S or $\neg$S. Since $\mathcal{A}$ is existentially complete in $\mathcal{B}$ there exist elements $\bar{a}'$ of $\mathcal{A}$ satisfying the condition $C(\bar{a},\bar{a}')$. Then $C(\bar{a},\bar{a}')$ is true in $\mathcal{A}$ and forces S or $\neg$S, as desired.

To check the inductivity of Σ^f again apply Proposition 45. If $\mathcal{A}$ is the union of a chain $\{\mathcal{A}_\alpha\}$ of finitely Σ-generic structures, and S is a sentence defined in $\mathcal{A}$, then for large α S is also defined in $\mathcal{A}_\alpha$, hence S or $\neg$S is forced by $\mathcal{A}_\alpha$, and hence S or $\neg$S is forced by $\mathcal{A}$.

<u>Proposition 48</u>. If $\mathcal{B}$ is in Σ^f and $\mathcal{A}$ is existentially complete in $\mathcal{B}$ then $\mathcal{A}$ is in Σ^f.

<u>Proof</u>: See the proof of Theorem 47.

<u>Corollary 49</u>. Let Σ be an inductive class of structures. Then Σ^f is cofinal in Σ iff $\Sigma^f = \Sigma^\infty$, in which case $\mathop{E}_\Sigma = \Sigma^f = \Sigma^\infty$.

<u>Proof</u>: Certainly if $\Sigma^f = \Sigma^\infty$ then Σ^f is cofinal in Σ. For the converse apply Lemma 20 and Theorem 47. **Finally, if** Σ^f is cofinal in Σ we apply Proposition 48 and Theorem 45 to conclude that Σ^f equals $\mathop{E}_\Sigma$.

The following result is an analog of Theorem 35.

<u>Theorem 50</u>. Let T be a theory in a countable language. Then Σ_T^f is complete iff Σ_T has the joint embedding property. (Definitions 32, 34).

<u>Proof</u>: Assume first that Σ is a class of structures with the joint embedding property and $\mathcal{A},\mathcal{B}$ are finitely Σ-generic. We will show that $\mathcal{A},\mathcal{B}$ are elementarily equivalent. Consider an arbitrary sentence S. Choose conditions C_1, C_2 in $\mathcal{A},\mathcal{B}$ respectively, each of which forces S or $\neg$S. Since $\mathcal{A},\mathcal{B}$ are contained in some element of Σ, it follows that $C_1 \cup C_2$ is a condition, and hence C_1, C_2 both force S or both force $\neg$S (by Lemma 42). Thus $\mathcal{A},\mathcal{B}$ both satisfy

S or both satisfy $\neg S$, as desired.

For the converse, assume Σ_T^f is complete and let $\mathcal{A}, \mathcal{B}$ be in Σ_T. We seek to jointly embed $\mathcal{A}, \mathcal{B}$ in a model of T, that is we seek a model of:

(*) $T \cup \mathrm{Diag}(\mathcal{A}) \cup \mathrm{Diag}(\mathcal{B})$.

It suffices therefore to show that (*) is a consistent theory.

Supposing the contrary, we obtain conditions C_1, C_2 true in $\mathcal{A}, \mathcal{B}$ respectively such that $T \cup C_1 \cup C_2$ is inconsistent; we may of course take the sets of constants occurring in C_1 and C_2 to be disjoint. We now apply Theorem 43 to obtain finitely Σ_T-generic models $\mathcal{A}', \mathcal{B}'$ in which C_1, C_2 are respectively true. Since $\mathcal{A}', \mathcal{B}'$ are elementarily equivalent, the statements:

1. "There are elements $\bar{a}$ satisfying the condition C_1."
2. "There are elements $\bar{b}$ satisfying the condition C_2."

are true in <u>both</u> $\mathcal{A}'$ and $\mathcal{B}'$. In particular $C_1 \cup C_2$ is consistent with T, a contradiction.

The following two theorems are of great intrinsic importance in the theory of finitely generic structures, and also can serve as the starting point for a very different approach to this theory.

<u>Definition 52</u>. Let T be a theory, $\mathcal{A}$ a model of T. $\mathcal{A}$ completes T iff for every model $\mathcal{B}$ of T extending $\mathcal{A}$, $\mathcal{A}$ is an elementary substructure of $\mathcal{B}$.

<u>Lemma 53</u>. Let T be a theory, $\mathcal{A}$ a completing model of T, S a sentence true in $\mathcal{A}$. Then for some condition C true in $\mathcal{A}$, $T \cup C$ proves S.

<u>Proof</u>: In fact one sees easily that the following are equivalent:

1. $\mathcal{A}$ satisfies S.
2. $T \cup \mathrm{Diag}(\mathcal{A})$ proves S.
3. $T \cup C$ proves S for some condition C true in $\mathcal{A}$. $\dashv$

<u>Theorem 54</u>. Let T be a theory in a countable language, T^{*f} the theory of Σ_T^f. Then the class of T^{*f}-complete models coincides with

Σ_T^f.

Proof: As a preliminary remark, consider a condition $C(\bar{a})$ which forces a sentence $S(\bar{a})$ relative to Σ_T. Then T^{*f} contains the sentence:

(*) $\forall \bar{x}\ (\bigwedge C(\bar{x}) \Rightarrow S(\bar{x}))$

(here of course $\bigwedge C$ is simply the conjunction of all formulas in C; it is evident that (*) holds in every model in Σ_T^f, as claimed).

Now let $\mathfrak{A}$ be in Σ_T^f, $\mathfrak{B}$ a model of T^{*f}, with $\mathfrak{A} \subseteq \mathfrak{B}$. We will show that $\mathfrak{A}$ is an elementary substructure of $\mathfrak{B}$. Fix any sentence $S(\bar{a})$ true in $\mathfrak{A}$, and let $C(\bar{a})$ be a condition forcing $S(\bar{a})$ and true in $\mathfrak{A}$. Then $C(\bar{a})$ holds in $\mathfrak{B}$, and since $\mathfrak{B}$ satisfies (*), $S(\bar{a})$ also holds in $\mathfrak{B}$.

Conversely let $\mathfrak{A}$ be a T^{*f}-complete model of T^{*f}. For any sentence S, if $\mathfrak{A}$ forces S then some condition C true in $\mathfrak{A}$ forces S; then (*) applies to show that $\mathfrak{A}$ satisfies S. We have thus proved half of the following assertion:

(**) For any sentence S defined in $\mathfrak{A}$, $\mathfrak{A}$ forces S iff $\mathfrak{A}$ satisfies S.

We now prove the other half by induction on the complexity of S, the only case of interest being $S = \neg S_1$. Assume therefore that $\mathfrak{A}$ satisfies S. We must show that $\mathfrak{A}$ forces S. By Lemma 53 we have a condition C true in $\mathfrak{A}$ such that $T^{*f} \cup C$ proves $\neg S_1$. Our claim is that C forces $\neg S_1$, and hence $\mathfrak{A}$ forces $\neg S_1$, as desired.

Suppose on the contrary that C does not force $\neg S_1$, i.e. C has an extension C' forcing S_1. Then by (*):

$T^{*f} \cup C'$ proves S_1.

Since $T^{*f} \cup C$ proves $\neg S_1$, $T^{*f} \cup C'$ is inconsistent. This contradicts Theorem 43.

Theorem 55. Let T be a theory in a countable language. Then there is exactly one theory T^* satisfying:

1. $\Sigma_{T^*} = \Sigma_T$.

2. T^* equals the theory of the class of T^*-complete models.

Proof: T^* exists: Let $T^* = T^{*f}$. Then 55.2 follows immediately from Theorem 53 and 55.1 follows easily from Theorem 43.

T^* is unique: Let T_1^*, T_2^* satisfy 55.1-2. We will show that T_1^*, T_2^* have the same completing models, and hence coincide. Thus we prove by induction on the complexity of sentences S:

(*) Let $\mathcal{A}$ be a completing model of T_1^*, $\mathcal{B}$ an extension of $\mathcal{A}$ satisfying T_2^*. Then $\mathcal{A}$ satisfies S iff $\mathcal{B}$ satisfies S.

The only case of interest involves sentences S of the form $\exists x\, S_1(x)$. Clearly if $\mathcal{A}$ satisfies such a sentence S then the induction hypothesis (*) applied to S_1 implies that $\mathcal{B}$ does as well. Assume therefore that $\mathcal{B}$ satisfies $\exists x\, S_1(x)$, whereas $\mathcal{A}$ does not. We set off in search of a contradiction.

Since $\mathcal{A}$ is a completing model of T_1^* there is a condition C true in $\mathcal{A}$ such that:

(A) $T_1^* \cup C$ proves $\neg\exists x\, S_1(x)$ (Lemma 53).

On the other hand $\mathcal{B}$ is a model of:

$T_2^* \cup C \cup \exists x\, S_1(x)$.

Using 55.2 (for T_2^*) we see that there is a completing model $\mathcal{B}'$ of T_2^* satisfying $C \cup \exists x\, S_1(x)$. In particular there is a condition C' true in $\mathcal{B}'$ and extending C such that:

(B) $T_2^* \cup C'$ proves $S_1(\bar{b})$ for certain constants $\bar{b}$ occurring in C'.

By 55.1 T_1^* has a model $\mathcal{A}'$ in which C' holds, and by 55.2 this model may be taken to be a completing model. Extend $\mathcal{A}'$ to a model $\mathcal{B}''$ of T_2^*. By (B):

$\mathcal{B}''$ satisfies $S_1(\bar{b})$.

By (A):

$\mathcal{A}'$ satisfies $\neg S_1(\bar{b})$.

This contradicts the induction hypothesis (*) applied to S_1.

§4. Existentially Complete Commutative Rings.

We are going to investigate the content of the preceding three sections in the case of the particular theory CR: the theory of commutative rings with identity. We will prove that CR is not companionable, and that $\Sigma_{CR}^{\infty} \cap \Sigma_{CR}^{f} = \emptyset$. More precisely, we will find a specific sentence called Reg which is true in all finitely generic commutative rings and false in all infinitely generic commutative rings. The statement Reg when interpreted in an existentially complete commutative ring A can be read as follows:

"A/rad(A) is von Neumann regular."

We mention in passing that Reg is an A_3-sentence (see Definition 58.3) and is not $\underset{\sim}{E}_{CR}$-persistent. Thus Theorem 13 is in general best possible.

Our results depend on three algebraic lemmas.

Lemma 56. Let a be an element of a commutative ring A. Then the following are equivalent:

1. a is not nilpotent.
2. In some extension B of A, a divides a nonzero idempotent.

Proof: If a divides the nonzero idempotent e in B, then for all n a^n divides $e^n = e$, and thus a is not nilpotent.

Assuming conversely that a is not nilpotent, we form the ring:

$B = A[x]/((ax)^2 - ax)$.

Polynomials p(x) over A may be construed both as elements of A[x] and of B. Thus for example if e = ax then e is an idempotent in B.

Evidently the canonical map A—>B is an embedding (the map B—>A induced by x->0 provides a left inverse). Since a divides the idempotent e in B our only concern is to show that $e \neq 0$ in B. This we do by explicit computation. Suppose on the contrary that in A[x] we can write:

$e = p(x)((ax)^2 - ax)$

for some polynomial $p(x) = \sum p_i x^i$. This produces a series of equations:

$$-a = ap_0,\quad a^2p_0 = ap_1,\ a^2p_1 = ap_2,\ a^2p_2 = ap_3,\dots .$$

In particular $-a^n = ap_{n-1} = 0$ for n large (since p_{n-1} is eventually zero). This of course contradicts the assumption that a is not nilpotent.

Corollary 57. CR is not companionable.

Proof: Suppose $\underset{\sim}{E}_{CR}$ is the class of models of a theory CR*. By Lemma 56, if we introduce a constant symbol a then:

$$CR^* \cup \{a^n \neq 0\colon \text{all } n\} \text{ proves "}a\text{ divides a nonzero idempotent".}$$

This implies that for sufficiently large n:

$$CR^* \cup \{a^n \neq 0\} \text{ proves " }a\text{ divides a nonzero idempotent,"}$$

and in particular $CR^* \cup \{a^n \neq 0\}$ proves "$a^{n+1} \neq 0$", which is impossible.

Definition 58.

1. If a is an element of a commutative ring A, a is regular iff a^2 divides a.

2. a is possibly nilpotent iff a divides no nonzero idempotent.

3. The statement Reg will be a formal version of: "Every element x is the sum of a regular element y_1 and a possibly nilpotent element y_2." Reg is written explicitly as follows:

$$\forall x \exists y_1 \exists y_2 \exists y_3 \forall z\ (x = y_1+y_2\ \&\ y_1^2y_3 = y_1\ \&((y_2z)^2 = y_2z \Rightarrow y_2z = 0))$$

Lemma 59. Let a be an element of the commutative ring A. Then the following are equivalent:

1. a^{n+1} divides a^n.
2. $a = a_1+r$ with a_1 regular, $r^n = 0$.
3. $a = a_1+r$ with a_1 regular, $r^n = 0$, $a_1r = 0$.

Proof: The argument being rather computational, we will leave a number of verifications to the reader.

1 => 2: If $a^{n+1}b = a^n$ then set $a_1 = a^nb^{n-1}$, $r = a-a_1$. Verify that $a = a_1+r$, $a_1^2b = a_1$, $r^n = 0$.

2 => 3: If $a = a_1 + r$, $a_1^2 b = a_1$, $r^n = 0$, then set

$a_2 = a_1(1+br)$, $s = r(1-a_1 b)$.

Then verify that $a = a_2 + s$, a_2 is regular, $s^n = 0$, $a_2 s = 0$.

3 => 1: If $a = a_1 + r$, $a_1^2 b = a_1$, $a_1 r = 0$, and $r^n = 0$ then $a^{n+1} b = a^n$.

Lemma 60. For a, b elements of a commutative ring A the following are equivalent:

1. a divides b in an extension of A.
2. For all x in A, if $ax = 0$ then $bx = 0$.

Proof: Clearly 1 => 2. For the converse, assume 2 and let

$B = A[x]/(ax-b)$.

It evidently suffices to verify that the canonical map $A \longrightarrow B$ is an embedding. We perform an explicit computation. Suppose c is an element of A and $c = 0$ in B, i.e. we have an equation in $A[x]$:

$c = (ax-b)p(x)$

for some polynomial $p(x) = \Sigma p_i x^i$. This produces a series of equations:

$-c = bp_o$, $ap_o = bp_1$, $ap_1 = bp_2$, ..., $ap_i = bp_{i+1}$, ...

For large n $bp_n = 0$. Whenever $bp_n = 0$ with $n > 0$ the above equations imply that $ap_{n-1} = 0$, and hence by condition 2 $bp_{n-1} = 0$. By downward induction it follows that $bp_o = 0$, and so $c = 0$. ⊣

Theorem 61. The statement Reg is false in all infinitely generic commutative rings, and true in all finitely generic commutative rings. In particular $\Sigma_{CR}^{\infty} \cap \Sigma_{CR}^{f} = \emptyset$.

Proof: We begin by considering an infinitely generic commutative ring A. We will find an infinitely generic extension B of A in which the failure of Reg is evident; since Σ_{CR}^{∞} is model complete it follows that Reg already fails in A.

In the first place one easily finds an extension A_1 of A containing elements $a, x_1, x_2, \ldots$ such that for each n:

$a^{n+1} x_n = 0 \neq a^n x_n$.

Then in any extension B of A_1, for each n a^{n+1} does not divide a^n. If B is infinitely generic, then since B is existentially complete it follows from Lemma 59 that Reg fails in B.

To see that Reg holds in any finitely generic commutative ring A, we fix an element a in A and suppose (toward a contradiction) that A forces:

$$(*)\ \neg\exists y_1\, \exists y_2\, \exists y_3\, \forall z\ (a = y_1+y_2 \,\&\, y_1^2 y_3 = y_1 \,\&\, ((y_2 z)^2 = y_2 z \Rightarrow y_2 z = 0)).$$

Let C be a finite subset of the diagram of A which forces (*). We will obtain a contradiction by finding an n such that

$$C \cup \{a^{n+1}c = a^n\}$$

is a condition. That this is a contradiction follows by comparing (*) with Lemma 59.

To find a suitable n, we let A_o be a finitely generated model of C (a suitable subring of A- generated by the constants occurring in C- will do). In particular A_o is noetherian. Let I_n be the annihilator of a^n in A_o. Since A_o is noetherian, the increasing chain of ideals $\{I_n\}$ is eventually constant, i.e. for large n we have: $I_n = I_{n+1}$. According to Lemma 60 a^{n+1} divides a^n in an extension of A_o (for n large). In particular $C \cup \{a^{n+1}c = a^n\}$ is a condition.

§5. Rings without Nilpotents.

The arguments of the preceding section naturally lead us to consider the theory of commutative rings without nilpotents, called SCR. Here we will prove the theorem of Carson-Lipshitz-Saradino: SCR is companionable (Theorem 72).

Definition 62. Let A be a commutative ring (with identity).

A is semiprime iff A contains no nilpotent elements,

regular iff every element of A is regular (Definition 58). (One sees easily that a regular nilpotent element must equal zero, and thus regular rings are semiprime.)

The theories of semiprime or regular commutative rings are denoted respectively SCR, RCR.

The main algebraic ingredients of this section will consist of some structure theory of regular rings, motivated largely by examples of the following sort.

Example 63. Let X be a compact totally disconnected Hausdorff space (totally disconnected = having a basis of clopen sets). Let F be a field carrying the discrete topology, and let $C(X;F)$ be the ring of locally constant functions from X to F. Then:

1. $C(X;F)$ is a regular ring.
2. The idempotent elements of $C(X;F)$ are exactly the characteristic functions of clopen sets.
3. The points of X are in 1-1 correspondence with the prime ideals I of $C(X;F)$ under the correspondences:

   $I_p = \{f \in C(X;F): f(p) = 0\}$.

   $p_I = \bigcap \{Z_f: f \in I\} \quad (Z_f = \{p: f(p) = 0\})$.
4. If p is a point of X and I_p is the corresponding prime ideal then the following diagram commutes:

$$\begin{array}{ccc} & C(X;F) & \\ {}^{\pi}\swarrow & & \searrow^{e_p} \\ C(X;F)/I_p & \xrightarrow{\ \simeq\ } & F \end{array}$$

   where π is the canonical projection and e_p is the evaluation map, $e_p(f) = f(p)$.

The verifications of points 1-4 are straightforward. To see for instance that $C(X;F)$ is regular, fix $f \in C(X;F)$ and define

$$f^{-1}(x) = \begin{cases} f(x)^{-1} & f(x) \neq 0 \\ 0 & f(x) = 0. \end{cases}$$

Then evidently $f^2 f^{-1} = f$, so f is regular.

Our immediate goal is to prove that every regular ring greatly resembles the rings of Example 63.

<u>Definition 64</u>. Let A be a commutative ring (with identity).

1. $\mathrm{rad}(A)$ is the set of nilpotent elements of A.

2. $S(A)$ is the set of prime ideals of A, carrying the topology generated by the following basis of open sets:

$O_a = \{p \in S(A)\colon a \notin p\}$ for a in A.

3. $B(A)$ is the set of idempotent elements of A, partially ordered by the divisibility relation:

$e \leq f$ iff f divides e.

("S" stands for "spectrum" or "Stone space" in 2; "B" stands for "Boolean" in 3- $B(A)$ is always a Boolean algebra, as we will verify below in a special case.)

We state two well known facts as a lemma, leaving the proof to the reader.

<u>Lemma 65</u>. Let A be a commutative ring with identity. Then:

1. $S(A)$ is compact.

2. $\mathrm{rad}(A) = \bigcap S(A)$ (the intersection of all prime ideals of A).

<u>Proof</u>: See the exercises.

<u>Definition 66</u>. Let A be a commutative ring with identity. The canonical map

$$A \longrightarrow \prod_{p \in S(A)} A/p$$

will be called the <u>canonical representation</u> of A. Its kernel is rad A by Lemma 65; thus the canonical representation of A is injective iff A is semiprime. For a in A, the residue of a in A/p will be denoted $a(p)$.

<u>Theorem 67</u>. Let A be a regular commutative ring. Then:

1. Under the canonical representation A is isomorphic with a ring of functions on $S(A)$, taking values in various fields A/p. Under this representation any idempotent e is identified with the characteristic function of O_e. (Cf. 64.2.)

2. $S(A)$ is a totally disconnected compact Hausdorff space, and a basis for the topology of $S(A)$ is given by the clopen sets

$\{O_e\colon e \in B(A)\}$.

The partially ordered set $B(A)$ is in fact a Boolean algebra (complemented distributive lattice) isomorphic to the Boolean algebra of all clopen subsets of $S(A)$; the Boolean operations on $B(A)$ may be expressed in terms of the ring operations as follows:

(B) $e \quad f = e+f-ef$, $e \cap f = ef$, $-e = 1-e$ (we write Boolean operations on the left and ring operations on the right here).

<u>Proof</u>:

1. Since A is semiprime, the canonical representation of A represents A as a ring of functions on $S(A)$ taking values in integral domains A/p. One verifies easily that each A/p is regular (since A is), and that every regular integral domain is a field.

For any idempotent e in A, $e(p)$ is an idempotent in A/p. Thus $e(p) = 0$ or 1, and $e(p) = 0$ iff $p \notin O_e$. Thus e is the characteristic function of O_e.

2. We note first that to any element a of A we may associate an idempotent e such that a divides e and e divides a. Simply choose b satisfying $a^2b = a$ and set $e = ab$.

Thus for any element a of A there is an associated idempotent e with $O_a = O_e$. In particular the topology on $S(A)$ is generated by the basis of open sets:

$\{O_e\colon e \in B(A)\}$.

Call this basis $B'(A)$. One sees easily that the map $e \longrightarrow O_e$ is an isomorphism between the partially ordered set $B(A)$ and the set $B'(A)$ partially ordered by $\subseteq$. Furthermore one sees easily that the Böolean operations $\cup, \cap, -$ defined by the equations (B) above correspond to set-theoretic $\cup, \cap, -$ in $B'(A)$. In particular $B'(A)$ is a Boolean algebra, $B(A)$ is isomorphic to $B'(A)$, $B(A)$ is a Boolean algebra, and the Boolean operations satisfy (B).

It remains to prove that $B'(A)$ consists of all clopen sets in $S(A)$ and that $S(A)$ is a totally disconnected compact Hausdorff space. Since the basis $B'(A)$ is closed under complementation, it consists of clopen sets. Thus $S(A)$ is totally disconnected, and compact according to Lemma 65. It follows easily that every clopen subset of $S(A)$ is a finite union of basic clopen subsets in $B'(A)$. Since $B'(A)$ is closed under unions, $B'(A)$ thus contains all clopen subsets of $S(A)$. We conclude by showing that $S(A)$ is Hausdorff.

Fix $p \neq q$ in $S(A)$, and choose a in $p-q$. Then q is in O_a and p is in the complement of O_a. By our previous remarks O_a is clopen, and it follows that $S(A)$ is Hausdorff. $\dashv$

Terminology 68. If A is a regular ring, we refer to elements p of $S(A)$ as points, and to elements a of A as functions. We identify each idempotent e in A with the characteristic function of O_e; we also identify e with the set O_e.

If $F(\bar{x})$ is a formula in the language of rings, p is a point in $S(A)$, and $\bar{a}$ are functions in A, we say that $F(\bar{a})$ is true at p iff A/p satisfies $F(\bar{a}(p))$. $F(\bar{a})$ is said to be true on an idempotent e iff $F(\bar{a})$ is true at every point of e. $F(\bar{a})$ is false on e iff $F(\bar{a})$ is false at every point of e. (Thus it is possible for a formula to be neither true nor false on a given idempotent.)

The basic model theoretic fact concerning regular rings is the following:

Theorem 69. Let A be a regular ring, $F(\bar{x})$ a first order quantifier-free formula in the language of rings, $\bar{a}$ functions in A. Then A contains the characteristic function of the set of points of $S(A)$ at which $F(\bar{a})$ is true; in other words there is an idempotent e such that $F(\bar{a})$ is true on e and false on $1-e$.

Proof: Suppose first that F is an atomic formula, $F(\bar{a}) =:$ "$p(\bar{a}) = 0$".

We saw in the proof of Theorem 67 that there is an idempotent e such that $0_{p(\bar{a})} = 0_e$. Then $F(\bar{a})$ is true on $1-e$ and false on e.

Now let $F(\bar{a})$ be a general quantifier-free formula, i.e. a Boolean combination of atomic formulas. We want to see that the set of points where $F(\bar{a})$ is true corresponds to an idempotent of A, in other words that this set is clopen in $S(A)$. Since we have verified this fact for atomic formulas it follows for general quantifier-free formulas (the clopen sets are closed under Boolean operations).

We are now in a position to study the model companion of SCR.

Definition 70. SCR* is the union of the following theories:

1. RCR.
2. "All monic polynomials of positive degree have a root."
3. "The Boolean algebra of idempotents contains no atoms."

Lemma 71. The class of models of SCR* is cofinal in Σ_{SCR}.

Proof: We may begin with a commutative semiprime ring A, which is to be extended to a model of SCR*. For this purpose it is convenient to have at hand a compact totally disconnected nonempty Hausdorff space X having no isolated points. (Such a space is called a "Cantor space".)

The canonical representation of A embeds A in a product of integral domains A/p, each of which may be embedded in some algebraically closed field F_p. Let $C_p = C(X;F_p)$ be the ring of locally constant functions from X to F_p. Identify F_p with the ring of constant functions in C_p. Let $B = \prod_{p\in S(A)} C_p$. We have embeddings:

$$A \longrightarrow \prod A/p \longrightarrow \prod F_p \longrightarrow \prod C_p = B.$$

Our claim is that B satisfies SCR*. One sees in the first place that SCR* is closed under products, so it suffices to verify that C_p satisfies SCR*; this is straightforward. We remark that the verification of axiom 70.2 is perhaps simplest if one notices that for every element c of C_p there is a finite partition $e_1,\ldots,e_k$ of X into clopen sets e_i such that c is constant on each e_i.

We can now prove the main result.

Theorem 72. SCR* is the model companion of SCR. In other words $\mathcal{E}_{SCR}$ coincides with the class of models of SCR*.

Proof: We must first show that every ring A in $\mathcal{E}_{SCR}$ satisfies SCR*. We simply apply Lemma 71 to embed A in a ring B satisfying SCR*. Then the definition of existential completeness and a glance at axioms 70.1-3 show that A satisfies SCR*.

The converse will require a substantial argument. Notice first that the content of axioms 70.1-3 is as follows:

70.1 "A may be represented as a ring of functions on a Stone space S(A) taking values in various fields A/p."

70.2 "Each field A/p is algebraically closed."

70.3 "S(A) has no isolated points."

(Axioms 70.1-3 say at least this much.)

Assume now that we have a model A of SCR* embedded in a semi-prime ring B which satisfies some existential assertion defined over A:

(E) $\exists \bar{x}\ q(\bar{x},\bar{a})$ (q is quantifier-free).

We want to show that (E) holds already in A.

As a preliminary remark we notice that q may be assumed to have the form:

(*) $\bigwedge_i r_i(\bar{x},\bar{a}) = 0\ \&\ \bigwedge_j s_j(\bar{x},\bar{a}) \neq 0$

for certain polynomials r_i, s_j. (In brief: q is certainly equivalent to a disjunction of formulas of the form (*), and the quantifier $\exists \bar{x}$ commutes with disjunction $\bigvee$.) Applying Lemma 71, we may also assume that B is also a model of SCR*, and is in particular regular.

Thinking of A as a ring of functions on S(A) we see that (*) is equivalent to:

1. For each i, $r_i(\bar{x},\bar{a})$ vanishes on S(A).

and 2. For each j, $s_j(\bar{x},\bar{a})$ does not vanish at some point of S(A).

This leads us to consider the following quantifier-free formulas:

$q_o(\bar{x})$ = "$\bigwedge_i r_i(\bar{x},\bar{a})=0$"

$q_j(\bar{x})$ = "$q_o(\bar{x})$ & $s_j(\bar{x},\bar{a}) \neq 0$."

Our main claim is the following: (*) is satisfiable in A iff:

(*1) $\exists\bar{x}\, q_o(\bar{x})$ is true at every point of S (= S(A)).

(*2) $\exists\bar{x}\, q_j(\bar{x})$ is true at some point of S (for each j).

It is trivial that the satisfiability of (*) implies the truth of (*1) and (*2). The converse depends on axiom 70.3.

Assume then that (*1), (*2) hold. For each j choose a point p_j such that $\exists\bar{x}\, q_j(\bar{x})$ is true at p_j, and choose $\bar{c}_j$ in A such that $q_j(\bar{c}_j,\bar{a})$ holds at p_j. Let e_j be the clopen set (idempotent) on which $q_j(\bar{c}_j,\bar{a})$ holds. Each e_j is nonzero; since B(A) is atomless we may assume that the sets e_j are nonempty and disjoint. Let e_o be the complement of $\bigcup_j e_j$. At each point p of e_o choose elements $\bar{c}_p$ of A such that $q_o(\bar{c}_p(p), \bar{a}(p))$ holds (use (*1)), and let e_p be the intersection of e_o with the clopen set on which $q_o(\bar{c}_p,\bar{a})$ holds. Then the cover $\{e_p: p \in e_o\}$ of e_o may be refined to a finite partition $\{e'_{p_i}\}$ of e_o. Define functions $\bar{c}$ on S(A) by:

$$\bar{c}(p) = \begin{cases} \bar{c}_{p_i}(p) & p \in e'_{p_i} \\ \bar{c}_j(p) & p \in e_j. \end{cases}$$

Then the $\bar{c}$ are in A; indeed $\bar{c} = \Sigma\bar{c}_{p_i}e'_{p_i} + \Sigma\bar{c}_j e_j$. Furthermore $q(\bar{c},\bar{a})$ holds, proving that (*) is satisfiable in A.

Thus the satisfiability of (*) in A (or in B) is equivalent to the truth of (*1) and (*2) in A (respectively, B). We must therefore prove the following:

if (*1), (*2) hold in B then (*1), (*2) hold in A.

It is not difficult to see that for any prime ideal p of A there is a point p' of S(B) such that $p' \cap A = p$ (see the exercises), and therefore for every p in S(A) we have at least one embedding:

(i) $A/p \longrightarrow B/p'$.

By axiom 70.2 A/p is existentially complete (cf. the exercises, §1). Since B satisfies (*1), (i) shows that A also satisfies (*1). Similarly to verify (*2), start with a point p' of $S(B)$ at which $\exists\bar{x}\ q_j(\bar{x},\bar{a})$ holds, set $p_j = p' \cap A$, and deduce from (i) and the existential completeness of A/p that $\exists\bar{x}\ q_j(\bar{x},\bar{a})$ holds at p_j.

Thus the proof is complete.

§6. A Generalized Nullstellensatz.

Let T be an inductive theory of commutative rings. We are going to define the "T-radical" of a polynomial ideal in such a way that the following theorem is true:

Theorem 73. Let A be an existentially complete model of T and let $p, p_1, \ldots, p_k$ be polynomials in $A[x_1, \ldots, x_n]$. Then the following are equivalent:

1. $V(p_1, \ldots, p_k) \subseteq V(p)$.
2. p is in the T-radical of the ideal $(p_1, \ldots, p_k)$.

Here of course $V(\bar{p})$ is the variety $\{\bar{a} \in A^n\colon p_1(\bar{a}) = p_2(\bar{a}) = \ldots = 0\}$.

Definition 74. Let T be a theory of rings, A a submodel of a model of T, I an ideal in $A[x_1, \ldots, x_n]$. Then the radical of I with respect to T is defined by:

$$T\text{-rad}(I) = \bigcap \{J\colon I \subseteq J,\quad A[\bar{x}]/J \text{ is embeddable in a model of } T, \text{ and } J \cap A = (0)\}.$$

The proof of Theorem 73 is an immediate consequence of the various definitions.

Proof of Theorem 73:

1 => 2: We assume $V(\bar{p}) \subseteq V(p)$, $I = (\bar{p})$, $I \subseteq J$, $J \cap A = (0)$, and $B = A[\bar{x}]/J$ is embeddable in a model of T. We must show that p is in J. If we assume on the contrary that p is not in J, then

the point $\bar{x} = (x_1, \ldots, x_n)$, viewed as a point in B^n, lies on $V(\bar{p})$ (since $I \subseteq J$) but not on $V(p)$ (since $p \notin J$). Consider then the sentence:

"There is a point on $V(\bar{p})$ which is not on $V(p)$"

This sentence is existential and is true in B. Thus it is true in A, contradicting 1.

<u>2 => 1</u>: Assume p is in the T-radical of $I = (p_1, \ldots, p_k)$ and that $\bar{a}$ is a point on $V(\bar{p})$. We will show that $\bar{a}$ is on $V(p)$.

Map $A[\bar{x}] \longrightarrow A$ via the evaluation map $e_{\bar{a}}(q) = q(\bar{a})$. Let J be the kernel of this map. Evidently $I \subseteq J$, $J \cap A = (0)$, and $A[\bar{x}]/J \simeq A$. It follows that $\text{T-rad}(I) \subseteq J$, and in particular $p \in J$, i.e. $e_a(p)$ is zero, and hence $\bar{a} \in V(p)$. ⊣

<u>Corollary 75</u>. Let A be an algebraically closed field and let $p, p_1, \ldots, p_k$ be polynomials in $A[x_1, \ldots, x_n]$. Then the following are equivalent:

1. $V(p_1, \ldots, p_k) \subseteq V(p)$.
2. For some integer n, $p^n \in (p_1, \ldots, p_k)$.

<u>Proof</u>: From the point of view of Theorem 73, the argument given in Chapter 1 §2 may be summarized as follows:

Step 1: A field is existentially complete iff it is algebraically closed.

Step 2: Let T be the theory of fields, I a polynomial ideal. Then $\text{T-rad}(I) = \{p: \text{for some } n \ p^n \in I\}$.

§7. <u>Notes</u>.

The general theory of existentially complete structures was developed over several decades by Abraham Robinson. Much of the terminology used here is relatively recent and slightly at variance with earlier usage.

The material of §2 may be developed using "infinite forcing," an analog of the finite forcing used in §3 (see |35|). On the other hand the use of finite forcing in §3 can be eliminated in favor of methods resembling those used in §2 (see [51]).|).

The treatment of existentially complete commutative rings in §4 comes from |56|, and our treatment of the semiprime case in §5 is based on |57|.

Structural questions concerning existentially complete models can be treated using stability theory [49].

Exercises.

§1.

1. Let Σ be an inductive class of structures. Show that $\mathbb{E}_\Sigma$ is cofinal with Σ. (As a first approximation, show that each structure $\mathfrak{A}$ in Σ has an extension $\mathfrak{A}'$ satisfying:

(*) If e is an existential sentence defined in $\mathfrak{A}$ and true in an extension of $\mathfrak{A}'$ in Σ, then e is true in $\mathfrak{A}'$.

Using (*) repeatedly, form the union $\overline{\mathfrak{A}}$ of a chain

$$\mathfrak{A} \subseteq \mathfrak{A}' \subseteq \mathfrak{A}'' \subseteq \cdots$$

and show that $\overline{\mathfrak{A}}$ is existentially complete.)

2. Let F be the theory of fields, OF the theory of ordered fields. Use the material of Chapter 1 §2 to prove that F and OF are companionable.

§2.

3. Let G be the theory of graphs $\langle V;E\rangle$ (E is a symmetric binary relation on the "vertices" V). Let G' be the theory of graphs having no cycles. Show that G' is not companionable but $\mathbb{E}_{G'} = \Sigma^{\infty}_{G'}$.

4. Let T be the theory of ordered sets $\langle S, < \rangle$ equipped with

infinitely many distinguished subsets P_i, subject to:

(Ax. i) If P_i is nonempty then $\text{card}(S) \leq i$.

Then T is a universal theory in a countable language. Show that T has finitely generic models of each finite size, but no infinite finitely generic models. Show that $\underset{\sim}{E}_T = \Sigma_T^\infty$.

5. Let $E, <$ be binary relation symbols, $\{P_i : i < \infty\}$ unary predicate symbols, $\{c_\alpha\}$ uncountably many constant symbols. Let T be the theory:

   1. "E is an equivalence relation."
   2. "$<$ linearly orders each equivalence class of E, and elements of distinct equivalence classes are incomparable relative to $<$".
   3. "The subset P_i has at most one element."
   4. "$\forall x\, P_i(x) \Rightarrow$ the equivalence class of x under E has at most i elements."

   Then T is a universal theory in a countable language. Show that $T \cup \{\neg Ec_\alpha c_\beta : \alpha \neq \beta\} \cup \{\neg P_i(c) : \text{all } i, c\}$ has no finitely generic models.

§4.

6. Show that an existentially complete commutative ring A satisfies Reg iff $A/\text{rad}(A)$ is regular (Definitions 62,64; Lemma 59 is helpful).

§5.

7. Show that $S(A)$ is compact.
8. Show $\text{rad}(A) = \bigcap S(A)$.
9. Show that the class of regular rings has the amalgamation property.
10. Let A be an existentially complete commutative ring, $A' = A/\text{rad}\, A$. Show that A' is in $\underset{\sim}{E}_{SCR}$ iff A satisfies Reg.

§6.

11. Describe the CR-radical and the SCR-radical of a polynomial ideal. Show that the analog of Theorem 73 fails for CR and SCR if the ideal $(p_1, \ldots, p_k)$ is replaced by a general polynomial ideal

I (not necessarily finitely generated).

IV. Existentially Complete Division Rings

Introduction.

DR will be the theory of division rings (skew fields). Axioms for DR are well known and will not be repeated here. The main object of study will be $\underset{\sim}{E}_{DR}$, the class of existentially complete division rings (Chapter III §1). Since the theory of existentially complete structures owes its existence to Hilbert's Nullstellensatz one may well ask what becomes of the Nullstellensatz in the present context. This is answered quite satisfactorily in §2.

In the notation of Chapter III the class of finitely generic division rings is denoted DR^f and the class of infinitely generic division rings is denoted DR^∞. When T is a companionable theory we know that $\underset{\sim}{E}_T = T^f = T^\infty$. To date DR is the least companionable theory known (an honor it shares with the theory of groups). In particular we will see in §3 that:

(*) $DR^f \cap DR^\infty = \emptyset$ and there are $2^{\aleph_0}$ different complete extensions of the theory $Th(\underset{\sim}{E}_{DR})$.

We will make a few comments on existentially complete groups in §4 (the same theorems apply with slightly simpler proofs). We include a theorem of Macintyre and Neumann concerning finitely generated subgroups of existentially complete groups.

The methods used in this chapter are a mixture of pure algebra and a little recursion theory, with model theory bridging the gap.

From the algebraic side we rely primarily on one fact of overwhelming importance for the present considerations:

Cohn's Theorem. The theory DR has the amalgamation property. More explicitly, any diagram of the following type can be completed:

(A) 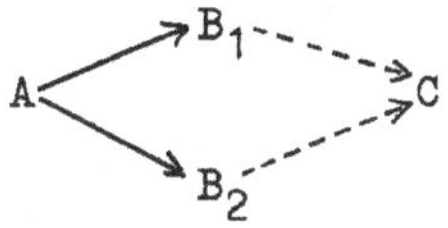

(we are in the category of division rings and embeddings).

The proof of this theorem **requires the generalization of various well understood aspects of commutative ring theory to a noncommutative setting. The most interesting fact to emerge from Cohn's proof of his theorem is the importance of his notion of the** **free** ideal ring, or fir (Definition 3), which apparently provides the "correct" generalization of the notion of principal ideal domain to a noncommutative setting.

We devote §1 to a lengthy sketch of the proof of Cohn's Theorem. In many cases the bare statement of the theorem suffices for model theoretic purposes, but see [33] for a case in which some of the details are relevant.

§1. Amalgamating Division Rings.

We will deal at some length with the ideas involved in the proof of:

Theorem 1. DR has the amalgamation property; i.e. any diagram:

(A) 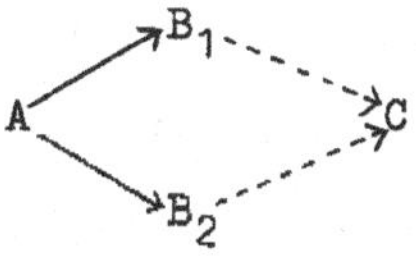

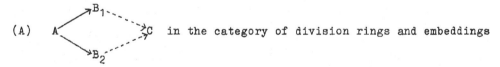

in the category of division rings and embeddings

can be completed as shown.

We begin with a brief synopsis of the method of **proof.**

Step 1.

Form the "universal product" $B_1 *_A B_2$ over A in the category of rings (free product with amalgamated subring). This is characterized up to isomorphism by the following universal diagram property:

(D) 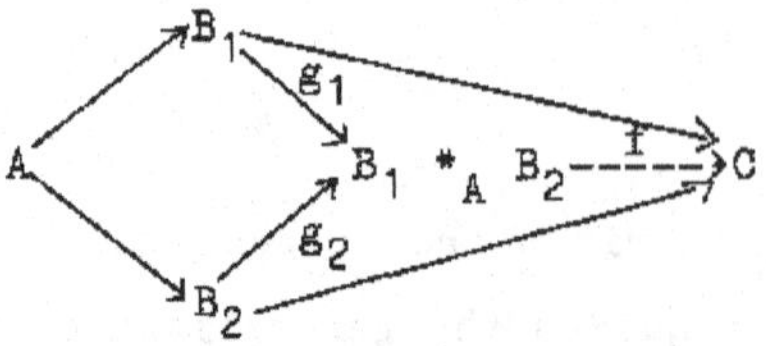

where C is arbitrary and we require that f is uniquely determined.

When A, B_1, B_2 are division rings it is particularly easy to give an explicit description of $B_1 *_A B_2$. In this case one can verify by inspection that each $g_i : B_i \longrightarrow B_1 *_A B_2$ is injective.

Let $R = B_1 *_A B_2$. If we embed R in a division ring C then we can easily complete diagram (A). In other words we can reformulate Theorem 1 as follows:

Theorem 2. Let $A \subseteq B_1, B_2$ be division rings, $R = B_1 *_A B_2$. Then R is embeddable in a division ring.

Step 2.

To prove Theorem 2 we need to make a study of embeddings of (noncommutative) rings into division rings. The commutative case has long been understood:

Fact. Let R be a commutative ring. Then R is embeddable in a field iff R has no zero divisors.

Viewed naively, this fact is completely misleading as far as the general case is concerned. For example:

Fact. There is a noncommutative ring R containing no zero divisors such that R admits no homomorphism into a division ring.

As it turns out, our ring $R = B_1 *_A B_2$ has the following two critical properties:

(I) Any left or right ideal of R is free as an R-module (in particular R has no zero divisors).

(II) Any two free R-modules of different dimensions are nonisomorphic.

Definition 3. Any ring satisfying I-II is called a fir (free ideal ring).

Obviously (I) is the key clause in the definition, but we will find ample use for (II) as well. (Compare Exercise 1.)

Our first step on the road to Theorem 2 is:

Theorem 4. Let $R = B_1 *_A B_2$ where $A \subseteq B_1, B_2$ are division rings. Then R is a fir.

Step 3.

The proof of Theorem 2 is completed by:

Theorem 5. Let R be a fir. Then R is embeddable in a division ring.

We will see that the key to a proof of Theorem 5 lies in the study of matrices over firs. We begin by reflecting on the following:

Fact. Let D be a division ring, T a square matrix over D. Then T is singular or invertible.

In the commutative case, a square matrix T is invertible as soon as its determinant $\det(T)$ is invertible. In particular, if R is a commutative ring then the process of embedding R into a field may be viewed in two ways:

(W1) Invert the nonzero elements of R.

(W2) Invert the square matrices T over R which are nonsingular (i.e. have nonzero determinant).

In the commutative context (W2) is a waste of breath. However in the noncommutative case (W1) is insufficient, whereas (W2) is adequate, if suitably interpreted. We have to supply a definition of "nonsingular"; there is more than one possibility |17|, but the following suffices for our purposes:

Definition 6. Let R be a noncommutative ring, T an $n \times n$ matrix over R. T is said to be nonfull ("singular") iff T can be written as the product:

$$T = A \cdot B$$

of an $n \times k$ matrix A and a $k \times n$ matrix B for some $k < n$. Otherwise T is <u>full</u> ("nonsingular").

To prove Theorem 5 one can in fact prove:

<u>Theorem 7</u>. Let R be a fir. Then R can be embedded in a ring S such that every full matrix T over R is invertible over S. Furthermore, any such S contains a division ring containing R.

The final statement bedomes somewhat plausible if one notes that all nonzero elements of R are full 1x1 matrices, and hence they at least are invertible in S.

For reasons that will become clear below, the following property of full matrices over firs is the key to Theorem 7:

<u>Definition 8</u>. A collection M of matrices over a ring R has the <u>diagonal sum property</u> (dsp) iff for any matrices A,B in M and any matrix C of suitable size:

(dsp) $\begin{pmatrix} A & C \\ 0 & B \end{pmatrix}$ is in M.

What does this correspond to in the commutative case? If R is a commutative ring and if M is defined as the set of all square matrices whose determinants lie in a fixed subset S of R, then the dsp amounts to:

S is a multiplicative subset of R.

Reflecting on the role of multiplicative sets in localization of commutative rings, it becomes plausible (though not yet clear) that the dsp should play a key role in the proof of Theorem 7.

Let us make this explicit:

<u>Theorem 9</u>. Let R be a fir, M the class of full matrices over R. Then M has the dsp.

With the help of the exercises we will give a reasonably complete account of certain aspects of the preceding material- notably Theorem 9 and the latter part of Theorem 7. Other aspects will be treated more superficially. In Step 1 there is little to prove, and we will be largely concerned with fixing notations and terminology.

The reader who is more interested in existentially complete division rings than in full matrices over firs will find this a good moment to jump to §2. We now take up a more detailed treatment of the following topics:

i. Structure of $B_1 *_A B_2$.

ii. Theorem 4.

iii. Theorem 7

iv. Theorem 9.

<u>Structure of</u> $B_1 *_A B_2$.

Let $A \subseteq B_1, B_2$ be rings. The ring $R = B_1 *_A B_2$ described in Step I above exists quite generally, but in general the "embeddings" $B_1 \longrightarrow R$, $B_2 \longrightarrow R$ need not be 1-1, and it may even happen that R is the zero ring. In our context A, B_1, B_2 are division rings, and there is no difficulty in giving a precise and explicit description of the ring R. (In the exercises we give a more polished version of the following, saying everything in terms of tensor products. This has its advantages.)

In the first place B_1, B_2 are vector spaces over A (we will emphasize the <u>right</u> vector space structure, for definiteness). Decompose:

$$B_1 = A \oplus B_1'$$

$$B_2 = A \oplus B_2'$$

and let u_i, v_j be bases over A for B_1', B_2' respectively. We refer to the u_i, v_j jointly as <u>basis elements</u>.

A <u>monomial</u> of height $n \geq 0$ is a formal expression:

(W) $\quad w_1 \ldots w_n a$

where a is in A and the basis elements $w_1, \ldots, w_n$ are alternately from B_1, B_2. The product $B_1 *_A B_2$ consists of all formal sums

(Σ) $\quad x = \sum_i m_i$ (finite sum),

where the m_i are monomials. We introduce the obvious operations

$+, \cdot$ acting on such formal sums (the definition of $\cdot$ is perhaps not quite obvious). Verification that $B_1 *_A B_2$ becomes a ring is carried out by routine manipulations, avoidable by using the terminology of Exercise 2.

The following notation and terminology is most useful. Finite sequences of the form (1212...) or (2121...) are called <u>types</u> and are denoted by capital letters I,J. A monomial (W) is of type I iff:

$w_i \in B'_{I(i)}$ for $i = 1,\ldots,n$.

B_I will denote the set of sums of monomials of type I. If I is the empty sequence (written: ()) let $B_{()} = A$. Notice that:

$B_{(1)} = B'_1$, $B_{(2)} = B'_2$, each B_I is a vector space over A.

$R = \oplus \sum_I B_I$.

Of course the maps $g_i: B_i \longrightarrow R$ are given by:

$B_i = A \oplus B'_i \simeq B_{()} \oplus B_{(i)} \subseteq R$. In particular g_1, g_2 are 1-1 and $B_1 \cap B_2 = A$.

We introduce some further terminology and notations. The length of a sequence I is denoted $|I|$. Let:

$H_n = \oplus \sum_{|I| \leq n} B_I$.

For x in $B_1 *_A B_2$, define the height

$h(x) = n$ iff $x \quad H_n - H_{n-1}$.

Notice that for each n there are two types of monomials of height n. Thus an element $x = \sum m_i$ of height n may contain either one or two types of monomials of height n; in the former case x is said to be <u>pure</u>, and in the latter case, <u>impure</u>.

It is evident that pure elements of height n are less complex than impure elements of height n, so we define a <u>modified height</u> function h' by:

$$h'(x) = \begin{cases} h(x) - 1/2 & \text{if } x \text{ is pure of positive height} \\ h(x) & \text{otherwise.} \end{cases}$$

Multiplication in R is reasonably well behaved if one pays close attention to the modified height and type of the elements involved. It is this fact which makes it possible to prove Theorem 4.

R <u>is a fir</u>.

$A \subseteq B_1, B_2$ are division rings, $R = B_1 *_A B_2$. We want to prove that R is a fir. We rely heavily on the notation of the preceding paragraphs.

Return to the commutative case in search of inspiration. Commutative firs are just principal ideal domains (Exercise 3). How does one prove a ring like $K[x]$ (with K a field) is a principal ideal domain? One introduces a degree function on $K[x]$, proves the Euclidean algorithm, and then applies this algorithm together with some form of induction on degree.

We can do much the same thing in the present context. We have degree functions h, h' on R. With sufficient ingenuity and patience one can formulate and prove a weak version of the desired algorithm, and use it to prove that R is a fir.

We will give an extremely sketchy account of this procedure, supplemented to some extent by the exercises.

We begin with a basic fact concerning height.

<u>Definition 11</u>.

1. Let r in R be pure of height n. If all monomials of height n occurring in r have type (12...) we say that r has <u>left type</u> 1. Otherwise r has left type 2. We define right type similarly.

2. An element r of R is <u>right reducible</u> iff for some invertible u in R, $h(ru) < h(r)$. Left reducibility is defined similarly.

We remark that every element of R is right associated to a right irreducible element ru.

Lemma 12 (Interaction Lemma). Suppose $a,b \in R$. Then:

1. $h(ab) = h(a) + h(b)$ unless a,b are pure and the right type of a coincides with the left type of b.

2. If a,b are pure and the right type of a equals the left type of b then:

   $h(ab) = h(a) + h(b) - 1$

   unless a is right reducible and b is left reducible.

Proof: (1) is essentially obvious. We leave (2) as Exercise 4.

Corollary 13. R has no zero divisors.

Proof: An easy consequence of the Interaction Lemma (Exercise 5).

We now come to the noncommutative Euclidean algorithm.

Theorem 14 (The Weak Algorithm). Let $\bar{a},\bar{b} \in R^n$. Suppose that for each i $h(a_i b_i) = N$ (constant), but $h(\Sigma a_i b_i) < N$. Fix a_j such that $h'(a_j) \geq h'(a_i)$ for all i. Then either:

1. Some a_i is right reducible.

or 2. There are c_i $(i \neq j)$ in R such that

   $h'(a_j - \sum_{i \neq j} a_i c_i) < h'(a_j)$.

For the proof see Lemma 3.4 of |13|; for a discussion of the Euclidean algorithm and noncommutative generalizations see |14|. We are more interested here in the fact that one can use the weak algorithm to prove that R is a fir. A sketch of the proof will be found in Exercises 6-8.

Embedding firs in division rings:

In the proof of the following result we see the importance of the diagonal sum property for full matrices over firs.

Theorem 15. Let R be a ring, M the set of full matrices over R, and S an extension of R such that every full matrix over R is invertible over S. If M has the diagonal sum property then R is contained in a division ring $K \subseteq S$.

<u>Proof</u>: Let K be the set of all first coordinates u_1 of n-tuples $\bar{u}$ in S^n such that for some full matrix T defined over R and for some vector $\bar{a}$ in R^n,

(*) $T\bar{u} + \bar{a} = 0$. (More suggestively, $u = T^{-1}(-\bar{a})$.)

We claim that K is a division ring containing R. It is trivial that:

1. $R \subseteq K$.

2. $-K \subseteq K$.

To see that:

3. $K + K \subseteq K$;

4. $K \cdot K \subseteq K$;

one passes from a pair of equations:

(**) $T\bar{u} + \bar{a} = 0, \; T'\bar{u}' + \bar{a}' = 0$

to the equations:

$$(+) \quad \begin{pmatrix} 1 & -1\ 0\ 0 \ldots 0 & -1\ 0\ 0 \ldots 0 \\ 0 & T & 0 \\ 0 & 0 & T' \end{pmatrix} \begin{pmatrix} u_1 + u_1' \\ \bar{u} \\ \bar{u}' \end{pmatrix} + \begin{pmatrix} 0 \\ a \\ a' \end{pmatrix} = 0.$$

$$(\cdot) \quad \begin{pmatrix} T & \bar{a}\ 0 \ldots 0 \\ 0 & T' \end{pmatrix} \begin{pmatrix} \bar{u}u_1' \\ \bar{u}' \end{pmatrix} + \begin{pmatrix} 0 \\ \bar{a}' \end{pmatrix} = 0.$$

Here we see the role of the dsp.

We have to work a little harder to prove:

5. $K^{-1} \subseteq K$.

We proceed as follows. Start with:

(*) $T\bar{u} + \bar{a} = 0$

and assume that $u_1 \neq 0$.

Let $T = (\bar{t}_1 \ldots \bar{t}_n)$ where $\bar{t}_i$ is the <u>ith</u> column of T. Let $\bar{v} = (1, u_2, \ldots, u_n)$ and $T' = (\bar{a}\ \bar{t}_2 \ldots \bar{t}_n)$. Then:

(*') $T'\bar{v} + \bar{t}_1 u_1 = 0$.

We will prove momentarily that T_1 is full. It then follows that T_1 is invertible over S, so that $\bar{v} = (-T_1^{-1}\bar{t})\,u_1$, and hence u_1 has a left inverse. It follows easily that u_1 is invertible (first show that the left inverse of u_1 has a left inverse), from which we derive:

(*") $T_1(\bar{v}u_1^{-1}) + \bar{t}_1 = 0$,

and hence u_1^{-1} is in K.

So it suffices to see that T_1 is full. Supposing the contrary, write $T_1 = A \cdot B$, where A is an nxk matrix and B is kxn, for some $k<n$.

Say $B = (\bar{b}\ B_1)$ where B_1 is $k\times(n-1)$. Then:

$$T = (\bar{t}_1\ AB_1) = (\bar{t}_1\ A)\begin{pmatrix}1 & 0\\ 0 & B_1\end{pmatrix};$$

since T is invertible over S, so is $(\bar{t}_1\ A)$. Thus if we write (*') in the form:

$$(\bar{t}_1\ A)\begin{pmatrix}u_1\\ \bar{v}\end{pmatrix} = 0$$

we may conclude that $u_1 = 0$, a contradiction.

Now we can proceed to embed any fir R in a ring S as required in the preceding theorem, and conclude that R is contained in a division ring. This was done by brute force in |15|. A better approach is suggested by the proof of Theorem 15.

According to the above proof of Theorem 15, we may think of the elements of K as having the form:

$$(T^{-1}\bar{a})_1$$

where T is full over R, $\bar{a} \in R^n$, and the subscript $_1$ means "first coordinate". This suggests that we <u>define</u> K as the set of <u>formal symbols</u>

$(\bar{a}/T)$ T full; $\bar{a}$ in R^n

modulo an appropriate equivalence relation and equipped with a suitably defined $+,\cdot$. This approach extends the usual construction of fields of fractions in a natural way. This method was successfully used in |17|, in conjunction with a simplifying trick outlined in Exercise 9.

The diagonal sum property.

We will now prove that full matrices over firs have the diagonal sum property. It might seem that some ingenious but superficial manipulation of full matrices is called for here. On the contrary, it is just at this point that we require a serious study of firs, and more particularly of finitely presented modules over firs.

Definition 16. Let M be an R-module (R is any ring).

M is finitely presented iff M is isomorphic with a quotient of finite dimensional free modules. In other words we require an exact sequence:

(P) $0 \longrightarrow R^m \xrightarrow{T} R^n \longrightarrow M \longrightarrow 0.$

Such a sequence is called a presentation of M. The homomorphism T may be thought of as an nxm matrix over R.

Conversely, given a matrix T which is not a right zero divisor then we define $M_T = R^n/TR^m$; M_T has the presentation (P). As we shall see, it will be convenient to study full matrices T in terms of the associated module M_T.

We will now undertake a lengthy detour through the theory of torsion modules over firs (Definition 19). It will turn out that the class of torsion modules coincides with the class of modules of the form M_T with T full. However the basic properties of torsion modules are better studied from a different point of view.

Definition 17. Let R be a ring, M a finitely presented R-module with presentation (P) as above.

1. R is said to have invariant basis number (ibn) iff any two free R-modules of unequal dimension are nonisomorphic. (This was clause II in the definition of firs.)

2. If R has ibn then we define:

the <u>rank</u> of M $(rk(M))$ is $m-n$ (see (P)).

For modules M which are not finitely presented we define:

$rk(M) = -\infty$ if M is finitely generated,

$rk(M) = +\infty$ if M is not finitely generated.

<u>Proposition 18</u>. If R has ibn then the function $rk(M)$ is well defined for all R-modules M.

<u>Proof</u>: This follows easily from Schanuel's Lemma (Exercise 10).

<u>Definition 19</u>. Suppose R has ibn and M is an R-module.

M is a <u>torsion</u> module iff:

1. $rk(M) = 0$.

2. $rk(N) \geq 0$ for $N \subseteq M$.

Let $Tn(R)$ be the category of torsion R-modules. In contemporary jargon, it can be shown that $Tn(R)$ is a noetherian artinian abelian category. We will be content with less. In fact the only theorem that interests us at the moment is the following, which will turn out to be equivalent to the dsp for full matrices over firs.

<u>Theorem 20</u>. Let R be a fir and suppose that

$$0 \longrightarrow M' \longrightarrow M \longrightarrow M'' \longrightarrow 0$$

is exact, with M', M'' torsion modules. Then M is a torsion module.

We will prove this shortly.

<u>Lemma 21</u>. Let R^n be a finitely generated free module over a fir R, and let $K \subseteq R^n$. Then K is free.

<u>Proof</u>: We proceed by induction on n, the case $n = 1$ being part of the definition of "fir". Now suppose the lemma is known for $n-1$, and let

$$K' = K \cap R^{n-1}.$$

Then $K/K' \simeq (K \oplus R^{n-1})/R^{n-1} \subseteq R^n/R^{n-1} \simeq R.$

Thus K/K' is isomorphic with a left ideal of R, and is therefore free. Then:

$0 \longrightarrow K' \longrightarrow K \longrightarrow K/K' \longrightarrow 0$

splits, i.e. $K \cong K' \oplus K/K'$. Since K' is free by the induction hypothesis, therefore K is free.

Lemma 22. Let $0 \longrightarrow M' \longrightarrow M \longrightarrow M'' \longrightarrow 0$ be an exact sequence of modules over a fir R.

1. If M', M'' are finitely generated then M is finitely generated and

$\mathrm{rk}(M) = \mathrm{rk}(M') + \mathrm{rk}(M'')$.

2. If M is finitely generated and M'' is finitely presented then M' is finitely generated.

Proof:

1. Form presentations:

$0 \longrightarrow K' \longrightarrow F' \longrightarrow M' \longrightarrow 0$

$0 \longrightarrow K'' \longrightarrow F'' \longrightarrow M'' \longrightarrow 0$

with F', F'' finitely generated and free. Then K, K' are free by Lemma 21. We may complete the following diagram:

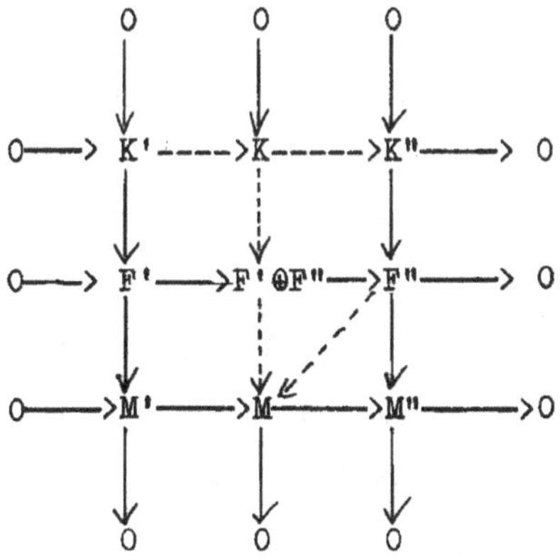

by diagram chasing, working from the bottom up (K is defined as the kernel of the corresponding map).

In particular M is finitely generated. Since K', K'' are free the top row splits and $K \cong K' \oplus K''$ is free.

In particular $rk(M) = \dim(F') + \dim(F'') - \dim(K') - \dim(K'')$

$$= rk(M') + rk(M'').$$

2. We complete the following diagram:

$$\begin{array}{ccccccccc}
 & & 0 & & 0 & & & & \\
 & & \downarrow & & | & & & & \\
 & & K & = & K & & & & \\
 & & \vdots & & \downarrow & & & & \\
0 & \longrightarrow & K' & \longrightarrow & F & \longrightarrow & M'' & \longrightarrow & 0 \\
 & & \vdots & & \downarrow & & & & \\
0 & \longrightarrow & M' & \longrightarrow & M & \longrightarrow & M'' & \longrightarrow & 0 \\
 & & \vdots & & \downarrow & & & & \\
 & & 0 & & 0 & & & &
\end{array}$$

(start by taking the middle column to be some presentation of M with F finite dimensional, free).

Then K' is free by Lemma 21 and since M'' is finitely presented it follows (from Schanuel's Lemma, Exercise 10) that K' is finite dimensional. Hence M' is finitely generated, as desired.

Proof of Theorem 20:

If $0 \to M' \to M \to M'' \to 0$ is an exact sequence and M', M'' are torsion modules, we are to prove that M is also a torsion module. M', M'' are in particular finitely generated, so Lemma 22.1 applies and we conclude that $rk(M) = rk(M') + rk(M'') = 0$.

To conclude the argument we must consider an arbitrary $N \subseteq M$ and verify that $rk(N) \geq 0$. We may take N to be finitely generated, since otherwise $rk(N) = \infty > 0$. We consider the exact sequence:

$0 \to N \cap M' \to N \to N/N \cap M' \to 0.$

Since $N/N \cap M' \xrightarrow{1-1} M/M' \simeq M''$,

$rk(N/N \cap M') \geq 0.$

By Lemma 22.2 it follows that $N \cap M'$ is finitely generated, so 22.1 applies:

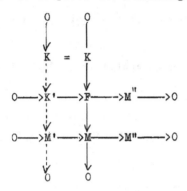

$$rk(N) = rk(N \cap M') + rk(N/N \cap M') \geq 0.$$

We will now relate torsion modules to full matrices.

Theorem 23. Let T be an $n\times n$ matrix, M an R-module such that

(P) $\quad 0 \longrightarrow R^n \xrightarrow{T} R^n \longrightarrow M \longrightarrow 0$

is exact. Then M is a torsion module iff T is full.

Remark. We supplement Theorem 23 by Exercise 11: If T is a full matrix over a fir, then (P) is exact.

Proof of Theorem 23:

Diagram chasing, laid out in Exercises 12-13.

We have one more diagram to chase:

Proof of Theorem 9 (dsp for full matrices over firs):

Consider two full matrices T' ($m\times m$) and T'' ($n\times n$) together with an $m\times n$ matrix A. Let $T = \begin{pmatrix} T' & A \\ 0 & T'' \end{pmatrix}$. We will show that T is full.

In the first place T remains a nonzero divisor (i.e. (P) above is exact). Put together the presentations corresponding to T', T, T'' in a commutative diagram:

$$\begin{array}{ccccccccc}
 & & 0 & & 0 & & 0 & & \\
 & & \downarrow & & \downarrow & & \downarrow & & \\
0 & \longrightarrow & R^m & \longrightarrow & R^m & \longrightarrow & M_{T'} & \longrightarrow & 0 \\
 & & \downarrow & & \downarrow & & \vdots & & \\
0 & \longrightarrow & R^{m+n} & \longrightarrow & R^{m+n} & \longrightarrow & M_T & \longrightarrow & 0 \\
 & & \downarrow & & \downarrow & & \vdots & & \\
0 & \longrightarrow & R^n & \longrightarrow & R^n & \longrightarrow & M_{T''} & \longrightarrow & 0 \\
 & & \downarrow & & \downarrow & & \downarrow & & \\
 & & 0 & & 0 & & 0 & &
\end{array}$$

After a short diagram chase, one sees that vertical maps may be inserted to make the diagram commute, and then the last column is automatically exact. By Theorem 20 M_T is a torsion module, so by Theorem 23 T is full.

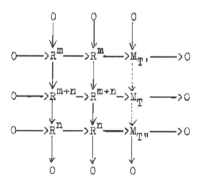

This concludes our sketch of the proof of Cohn's Theorem.

§2. Existentially Complete Division Rings- Algebraic Aspects.

Notation.

1. When D is a division ring, X a subset of D let $\langle\langle X\rangle\rangle$ denote the division ring generated by X in D.

2. If D is a division ring let $D[\bar{x}]$ be the usual polynomial ring in indeterminates $\bar{x}$ over D (the indeterminates $\bar{x}$ are taken to commute with one another and with D).

The most important application of the amalgamation property is the following:

Theorem 24. Let $A,B \subseteq D$ be division rings and suppose A, B are isomorphic via

$$h: A \simeq B.$$

Then there is a division ring D' extending D and containing an element t such that

$$t^{-1}at = h(a) \quad \text{for } a \text{ in } A.$$

In other words, every isomorphism between substructures of D is potentially the restriction of an inner automorphism.

Proof: We sketch a proof and leave much of the work to the reader in Exercises 14-16.

First we will extend D to a division ring D' with the following property:

(*) h can be extended to an automorphism h' of D'.

Consider in the first place a particular element d of D, and as a first step toward (*) let us attempt to extend h to the division ring $\langle\langle A,d\rangle\rangle$. In other words we seek an element d' of D such that $\langle\langle A,d\rangle\rangle \simeq \langle\langle B, d'\rangle\rangle$. There may well be no such element d' in D, but we can always extend D by adjoining a suitable d'. Just amalgamate:

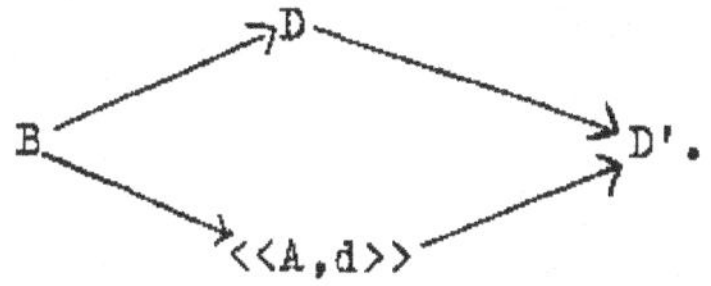

By iterating this procedure as often as necessary, one arrives at (*) (Exercise 14).

Simplifying the notation, we again write h for h', D for D'; but now h is an automorphism of D. We want to extend D to a division ring D' in which h extends to an inner automorphism.

We begin by formally adjoining an element t to D such that $t^{-1}at = ha$ for a in D. This is done by forming a <u>skew polynomial ring</u> $D[t;h]$ which is just the ordinary polynomial ring $D[t]$ equipped with a new multiplication. As usual, we write elements of $D[t;h]$ in the form:

$$p(t) = \Sigma a_i t^i \qquad (a_i \text{ in } D).$$

The multiplication is uniquely determined by associativity, distributivity, the rule $t^i t^j = t^{i+j}$, and the rule:

$$ta = h^{-1}(a)t.$$

To complete the argument we must somehow embed $D[t;h]$ in a skew field. This turns out to be reasonably easy by direct methods (Exercises 15-16). This concludes our sketch.

<u>Corollary 25</u>. Let D be an existentially complete division ring, $a_1,\ldots,a_n$, $b_1,\ldots,b_n$ in D, and suppose $\langle\langle\bar{a}\rangle\rangle \simeq \langle\langle\bar{b}\rangle\rangle$ via an isomorphism taking a_i to b_i. Then there is an inner automorphism of D taking a_i to b_i.

<u>Proof</u>: By the preceding theorem there is an element t in an extension D' of D such that $t^{-1}a_i t = b_i$ for all i. Since D is existentially domplete, there is also such an element in D.

<u>Notation</u>. Let D be a division ring.

1. For any subset X of D, the centralizer of X in D is denoted $C(X)$. $CC(X)$ is denoted $C^2(X)$.

2. For a,t in D let $a^t = t^{-1}at$.

<u>Theorem 26</u>. Let D be an existentially complete division ring, $a_1,\dots,a_n \in D$. Then:

1. $C^2(\bar{a}) = \langle\langle \bar{a} \rangle\rangle$.

2. The center of D is the prime field $\langle\langle 1 \rangle\rangle$ (a proper subfield of D).

<u>Proof</u>: Let $A = \langle\langle \bar{a} \rangle\rangle$.

1. Clearly $A \subseteq C^2(\bar{a})$. On the other hand if b is in $C^2(\bar{a})$ then D satisfies:

(*) $\forall x$ (if x commutes with $a_1,\dots,a_n$ then x commutes with b).

It follows that the statement (*) holds in every extension of D (since its negation is existential and D is existentially complete). Consider in particular a division ring D' containing the amalgamated product of D and $A[x]$ over A. Evidently the only elements of D which commute with x in D' are the elements of A. Thus by (*) b is in A; in short $C^2(\bar{a}) \subseteq A = \langle\langle \bar{a} \rangle\rangle$, as desired.

2. This is a special case of 1. (The final remark says that D is not commutative. This is obvious.)

<u>Theorem 27</u>. Any countable D in $\underset{\sim}{E}_{DR}$ has $2^{\aleph_0}$ automorphisms.

<u>Proof</u>: Write $D = \bigcup D_n$ as a union of finitely generated division rings. We will call any isomorphism h between finitely generated division rings contained in D a <u>partial automorphism</u>. This terminology is justified by Corollary 25. We call two partial automorphisms h_1,h_2 <u>compatible</u> if they are contained in a common extension h_3, and incompatible otherwise. We claim:

(*) any partial automorphism h has two incompatible extensions h', h".

(To complete the proof of the theorem we use the canonical argument for proving the existence of $2^{\aleph_0}$ objects of any desired kind. Namely we build a binary branching tree of height $\aleph_0$, placing a partial automorphism h at each node of the tree, in such a way that:

i. Each h is followed by two incompatible h', h''.

ii. At level n of the tree, each partial automorphism has domain and range containing D_n.

Then any one of the $2^{\aleph_0}$ distinct paths p through the tree gives rise to an automorphism h_p of D.)

To prove (*), let A be the domain of h. Choose a finitely generated division ring $B \subseteq D$ which properly contains A. By Corollary 25 h extends to an inner automorphism h' of D. The restriction of h' to B provides an extension of h. We will modify h' slightly to obtain a second, incompatible extension.

Since B is strictly larger than A, there is according to Theorem 26.1 an element of B which is not in $C^2(A) = A$. Let b be such an element and choose an element d of D which commutes with A but not with b. In other words the inner automorphism determined by d fixes A but moves b.

Define $h''(x) = h'(x^d)$ for x in B. Then h'' is a partial automorphism extending h which is incompatible with h'.

Let us consider now how we may reasonably generalize the notions of algebraic geometry to a noncommutative setting (cf. Chapter I §2). If D is a division ring we will want to replace the polynomial ring $D[\bar{x}]$ by a <u>noncommuting polynomial ring</u> $D\langle\bar{x}\rangle$. This may be defined as the free product with amalgamated subring $D *_F F[\bar{x}]$, where

$F = \langle\langle 1\rangle\rangle \subseteq D$

is the prime subfield. $D\langle x\rangle$ contains such polynomials as

(p) $ax-xa$ (a in D);

and for a in $D-F$, such polynomials p are nonzero.

Given a polynomial $p(\bar{x})$ in $D\langle\bar{x}\rangle$ and elements $\bar{a}$ of D, the element $p(\bar{a})$ in D is well defined, and the map:

$e_{\bar{a}}: D\langle\bar{x}\rangle \longrightarrow D$ defined by $p(\bar{x}) \longmapsto p(\bar{a})$ (evaluation at $\bar{a}$)

is a homomorphism (which would not be the case if we were so careless as to use the ordinary polynomial ring $D[\bar{x}]$).

Given polynomials $p_1(\bar{x}),\ldots,p_k(\bar{x})$ we define the variety

$V(p_1,\ldots,p_k) \subseteq D^n$

as the set of common zeros of $p_1,\ldots,p_k$ in D^n. More generally, if I is an ideal in $D\langle\bar{x}\rangle$ we define $V(I)$ as the set of common zeros of all p in I;

$V(I) \subseteq D^n$.

Since $D\langle\bar{x}\rangle$ need not be noetherian, not all varieties need be determined by finitely many polynomials. Example 30 below provides a striking illustration of this remark.

The reader who has taken §1 to heart may feel that the foregoing definition of variety mimics the commutative case a bit too faithfully. It would be closer to the spirit of §1 to consider more general varieties defined as the set of points in D^n where a given set of matrices of polynomials become nonfull. We will be content with the notion of variety introduced above, but a more courageous view of algebraic geometry is championed in |17|.

What becomes of the Nullstellensatz (Chapter 1, §2) in this setting? The ordinary Hilbert Nullstellensatz may be phrased as follows:

<u>Fact</u>. (Hilbert Nullstellensatz). For F an algebraically closed commutative field and for $p,p_1,\ldots,p_k$ in the polynomial ring $F[\bar{x}]$ the following are equivalent:

1. $V(p) \supseteq V(p_1,\ldots,p_k)$.
2. $p \in \text{radical}(p_1,\ldots,p_k)$.

Here the <u>radical</u> of an ideal I in a commutative ring R is defined equivalently by any of the following conditions:

i. $\text{radical}(I) = \{x\colon \text{For some } n \quad x^n \in I\}$.

ii. $\text{radical}(I) = \bigcap\{J\colon I \subseteq J \text{ and } J \text{ is prime}\}$.

iii. $\text{radical}(I) = \bigcap\{J\colon I \subseteq J,\ R/J \text{ is embeddable in a field}\}$.

A similar theorem holds in our context if we define:

Definition 28.

radical(I) = $\bigcap \{J: I \subseteq J,\ R/J$ is embeddable in a division ring$\}$.

Theorem 29. Let D be an existentially complete division ring, and let $p, p_1, \ldots, p_k \in D\langle\bar{x}\rangle$. Then the following are equivalent:

1. $V(p) \supseteq V(p_1, \ldots, p_k)$.
2. $p \in \text{radical}(p_1, \ldots, p_k)$.

Proof: When the definitions involved are unraveled it turns out that we have defined the class of existentially complete rings as those for which 1,2 are equivalent.

1 => 2. Fix an ideal J containing $(p_1, \ldots, p_k)$ in $R = D\langle\bar{x}\rangle$ and suppose that R/J is embeddable in a division ring. We must verify that p is in J. Look at the maps:

$$D \longrightarrow R \xrightarrow{\pi} R/J \longrightarrow D'.$$

We may view p as an element of R, R/J, or D'. We may also view p as a function on D and D'. The same remarks apply to the polynomials $p_1, \ldots, p_k$ and to the indeterminates $x_1, \ldots, x_n$.

Consider the point $\bar{x} = (x_1, \ldots, x_n)$ in $(D')^n$. This lies on the variety $V(p_1, \ldots, p_k)$, since if we evaluate the function p_i at $\bar{x}$ we just get p_i, which is 0 in D' (since p_i is in $J = \ker \pi$). Is $\bar{x}$ in $V(p)$ as well?

I. If $\bar{x}$ is in $V(p)$ then $p = p(\bar{x}) = 0$ in D', i.e. p is in J, as desired.

II. If $\bar{x}$ is not in $V(p)$ then D' satisfies:

(*) $\exists \bar{x}$ $\bar{x}$ is in $V(p_1, \ldots, p_k)$ but not in $V(p)$.

Since D is existentially complete (*) also holds in D, violating our hypothesis 1.

2 => 1. Let $\bar{a}$ in D^n be a point on $V(p_1, \ldots, p_k)$. Let $e = e_{\bar{a}}: D\langle\bar{x}\rangle \longrightarrow D$ be the evaluation map at $\bar{a}$. Let $J = \ker(e)$.

Then $(p_1, \ldots, p_k) \subseteq J$ and $R/J \simeq D$. Hence by assumption 2, p is in J, i.e. $e(p) = 0$. This means that $\bar{a} \in V(p)$. It follows that $V(p_1,\dots,p_k) \subseteq V(p)$, as desired.

This Nullstellensatz, though formally identical with the commutative Nullstellensatz, is less definitive than the latter. In the commutative case all varieties are determined by finitely many polynomials, so the Nullstellensatz is quite general. Over skew fields the question of extending the Nullstellensatz to varieties determined by infinitely many polynomials arises. However the following is a counterexample.

Example 30. Let D be a countable existentially complete division ring and let h be an outer (i.e. not inner) automorphism of D. Let I be the ideal generated by the following polynomials:

(p_a) $ax - xh(a)$ (a in D).

Then the polynomial x does not lie in the radical of I, but $V(I) \subseteq V(x)$.

We leave the proof as an excellent exercise for the reader (Exercise 17).

The following is perhaps the most surprising result concerning the structure of countable existentially complete division rings.

Theorem 31. If D is a countable existentially complete division ring then D contains a proper subring isomorphic with D.

Proof: Fix an element of D not in <<1>>. We first seek a division ring $D' \subseteq D$ such that:

1. d is not in D'.
2. Every finitely generated division ring contained in D is isomorphic with a subring of D'.
3. D' is existentially complete.

We will then supply a fairly routine argument that proves that D and D' are isomorphic.

Write $D = \bigcup A_n$ as a union of finitely generated division rings. We will find an increasing sequence $\{B_n\}$ of finitely generated division rings contained in D such that:

1'. d is not in any B_n.

2'. A_n is isomorphic with a substructure of B_n.

Then setting $D' = \bigcup B_n$, we see that 1,2 hold.

We define the B_n inductively, taking $B_{-1} = \langle\langle 1 \rangle\rangle$. Having found B_{n-1}, we consider the following assertion:

(*) For some t, d is not in $\langle\langle B_{n-1}, A_n^t \rangle\rangle$.

We will show that (*) is true in D, and therefore we may take $B_n = \langle\langle B_{n-1}, A_n^t \rangle\rangle$ for suitable t. Let us replace (*) by

(**) For some t,u: u commutes with $\langle\langle B_{n-1}, A_n^t \rangle\rangle$ but not with d.

By Theorem 26 (**) is equivalent to (*). To show that (**) holds in D it suffices to find an extension D'' of D in which (**) holds (since (**) is an existential assertion).

As a first approximation, let D'' be obtained by amalgamating the division ring $\langle\langle B_{n-1}, A_n \rangle\rangle(u)$ (see Exercise 16) with D over B_{n-1}. We denote the new copy of A_n by A_n' to avoid confusion.

Now u commutes with A_n' and B_{n-1}, but not with d. To obtain (**) we apply Theorem 24 to obtain a further extension of D'' in which A_n' is conjugate to A_n via an inner automorphism determined by some element t. Then clearly (**) holds, and this completes the construction of B_n.

We set $D' = \bigcup B_n$. As we have remarked, D' satisfies 1,2 above. We need only check that $D' \simeq D$. This is done by constructing a sequence of partial isomorphisms h_n between D' and D satisfying:

3. The domain D_n of h_n is a finitely generated division ring and $A_n \subseteq D_n \subseteq D$.
4. The range R_n of h_n is a finitely generated division ring and $B_n \subseteq R_n \subseteq D'$.
5. $h_n: D_n \simeq R_n$.

We choose h_n inductively, We will only worry about conditions 3,5. (By a symmetrical argument one may also ensure 4.) Suppose then that we have constructed:

$h_{n-1}: D_{n-1} \simeq R_{n-1}$.

Let $D_n = \langle\langle D_{n-1}, A_n \rangle\rangle$. There is a copy D_n' of D_n in D', by 2 above. Let $h: D_n \simeq D_n'$. Then we may apply Corollary 25 to:

(i) $h[D_{n-1}] \simeq h_{n-1}[D_{n-1}]$

to deduce that the isomorphism (i) may be taken to be the restriction to $h[D_{n-1}]$ of an inner automorphism f. Setting $h_n = fh$, we see that h_n extends h_{n-1} and satisfies conditions 3,5, as desired.

§3. Existentially Complete Division Rings-Model Theoretic Aspects.

We propose to show in this section that DR is not companionable, and that indeed the hierarchy DR^n is nondegenerate, no finitely generic division ring is infinitely generic, and there are $2^{\aleph_0}$ different complete theories of existentially complete division rings. As in the preceding section the main algebraic tool is the amalgamation property. Note that the results of this section apply equally well to DR_p (the theory of division rings of fixed characteristic) and are more striking in this form (Exercise 18).

Theorem 32. The class $\underset{\sim}{E}_{DR}$ of existentially complete division rings is not first order axiomatizable.

Proof: Consider two elements t_1, t_2 of a division ring D in $\underset{\sim}{E}_{DR}$, and suppose that t_1, t_2 are transcendental over $\langle\langle 1 \rangle\rangle$. Then by Corollary 25

(*) there is an x in D such that $x^{-1}t_1x = t_2$.

Thus if $\underset{\sim}{E}_{DR}$ is the class of models of some first order theory T, and if we introduce two constant symbols t_1, t_2, then:

(**) $T \cup \{p(t_1) \neq 0: p \in \langle\langle 1 \rangle\rangle[x]\} \cup \{p(t_2) \neq 0: p \in \langle\langle 1 \rangle\rangle[x]\}$ proves (*) above.

But clearly no finite subset of the theory in (**) proves (*). This is a contradiction (Chapter 0 §2).

Our main objective in the rest of this section is to show that division rings in DR^∞ are extremely complicated objects. The next theorem is the one important result in this connection. Its meaning will be clarified somewhat by Definition 34 and the succeeding discussion, and a more precise version is formulated in Definition 35 and Theorem 36. Using the machinery of recursion theory we will see at the end of this section that Theorem 33 (or better, Theorem 36) implies the results stated at the beginning of this section.

Theorem 33.

1. First order arithmetic is interpretable in $\underset{\sim}{E}_{DR}$.

2. Second order arithmetic is interpretable in DR^∞.

Definition 34. Let N be the set of natural numbers $\{0,1,\ldots\}$ and let P be the collection of all subsets of N.

1. The following structure is called "standard first order arithmetic" and will be denoted ${}^1N$:

$\langle N;+,\cdot\rangle$.

The theory $Th({}^1N)$ is called arithmetic (or complete arithmetic). There are of course innumerable models of arithmetic other than $Th({}^1N)$.

2. The following structure is called "standard second order arithmetic" (or "analysis") and will be denoted ${}^2N$:

$\langle N,P;\ +,\cdot,\in\rangle$. (P = power set of N.)

Here $\in$ is the membership relation "$n \in S$" between integers n and sets of integers S.

Oddly enough ${}^1N$ and ${}^2N$ considered as algebraic structures are of interest in connection with division rings in $\underset{\sim}{E}_{DR}$ and DR^∞ respectively. Before explaining this connection we pause to remind the reader of some well known algebraic phenomena. Consider the following types of algebraic structures:

1. Boolean algebras and commutative rings.

2. Division rings and desarguesian projective planes.

Discussion.

1. If R is a commutative ring then the set $B(R)$ of idempotent elements of R has a natural Boolean algebra structure (Chapter III §4). Notice that the set $B(R)$ is trivially first order definable in terms of the language of rings, and the Boolean operations are similarly definable. The structure of $B(R)$ carries a certain amount of information about R- namely information concerning the ways R can be decomposed into a direct product of other rings.

2. If P is a desarguesian projective plane then we can associate to P a division ring $D = D(P)$. The definition of D involves the use of certain parameters from P; but the isomorphism type of D is independent of the particular parameters used. The set D and the operations $+,-,\cdot,^{-1}$ on D are definable from the parameters used to construct D.

(Just as the category-theoretical algebraist is obsessed by functoriality, the model-theoretic algebraist is obsessed by definability.)

In a similar fashion we will construct:

3. models of arithmetic $^1N(D)$ in existentially complete division rings D.

4. models of second order arithmetic $^2N(D)$ in infinitely generic division rings D.

The constructions of $^1N(D)$, $^2N(D)$ will depend (in a uniform way) on finitely many parameters from D, and the operations and relations on $^1N(D)$, $^2N(D)$ will be definable in the language of division rings.

Definition 35. Let D be a division ring, $s,t \in D$. Define:

1. $N_t = \{t^n : n \geq 0\}$.
2. $t^m \oplus_t t^n = t^{m+n}$.

3. $t^m \otimes_t t^n = t^{mn}$.

4. $^1N_t = \langle N_t;\ \oplus_t,\ \otimes_t\rangle$.

5. For a in D let $S_{st}(a) = \{x \in N_t:\ s^{xa} = s^x\}$ (we continue to use the exponential notation for inner automorphisms).

6. For a in D and x in N_t set:

   $Elt_{st}(x,a)$ iff x is an element of $S_{st}(a)$.

7. For a,b in D set: $a =_{st} b$ iff $S_{st}(a) = S_{st}(b)$.

8. $^2N_{st} = \langle N_t, D;\ \oplus_t,\ \otimes_t,\ Elt_{st},\ =_{st}\rangle$.

(We may replace D in $^2N_{st}$ by the collection of equivalence classes of D modulo $=_{st}$.)

If t is of infinite order then ^1N_t is isomorphic with ^1N. Normally this fact is utterly devoid of interest, but we will see below that for existentially complete D, the set N_t and the operations $\oplus_t$, $\otimes_t$ are definable in the language of division rings. This is obvious for $\oplus_t$, surprising for N_t, and remarkable for $\otimes_t$. On the other hand it is obvious that the additional structure imposed on $^2N_{st}$ is definable in the language of division rings, but it is far from obvious that $^2N_{st}$ ever resembles ^2N.

All the algebraic results of this section will be variants on the theme of Theorem 26.1. There it was seen that for any finite set of parameters $\bar{a}$ in an existentially complete division ring, the set $\langle\langle\bar{a}\rangle\rangle$ is first order definable.

We now give a precise version of Theorem 33. The clauses of interest are II 1-5 and III2.

__Theorem 36.__ Let s,t be elements of an existentially complete division ring D.

I. The following predicates are definable:

1. "x is transcendental over $\langle\langle 1\rangle\rangle$"

2. "The elements $\{y^{(x^n)}: n \geq 0\}$ are algebraically independent over $\langle\langle 1\rangle\rangle$ and commute with one another."

II. Assume that t is transcendental over $\langle\langle 1\rangle\rangle$ and that the set $\{s^{(t^n)}: n \geq 0\}$ is a commuting family of algebraically independent elements over $\langle\langle 1\rangle\rangle$. Then the following predicates are definable:

1. "$x \in N_t$".
2. "$x \oplus_t y = z$".
3. "$x \otimes_t y = z$".
4. "$\mathrm{Elt}_{st}(x,y)$".
5. "$x =_{st} y$".

III. Under the assumptions of II,

1. ${}^1N_t \simeq {}^1N$.
2. If $D \in DR^{\infty}$ then ${}^2N_{st}$ is a model of second order arithmetic.

IV.

1. D contains at least one element t transcendental over $\langle\langle 1\rangle\rangle$.
2. If $D \in DR^{\infty}$ then D contains at least one pair s,t such that $\{s^{(t^n)}: n \geq 0\}$ is a commuting family of algebraically independent elements over $\langle\langle 1\rangle\rangle$.

The proof of this theorem is contained in the following Lemmas (37-41).

<u>Convention</u>. Call a pair of elements s,t in a division ring D <u>suitable</u> iff $\{s^{(t^n)}: n \geq 0\}$ is a commuting family of algebraically independent elements. Until the end of Lemma 41 D will be a fixed division ring which is (at least) existentially complete.

<u>Lemma 37</u>. For x in D the following are equivalent:

1. x is transcendental over $\langle\langle 1\rangle\rangle$.
2. $\langle\langle x\rangle\rangle$ is isomorphic with, but unequal to, $\langle\langle x^2\rangle\rangle$ (via a map carrying x to x^2).

<u>Proof</u>: Evidently 1 => 2. For the converse notice that 2 implies that $\langle\langle x\rangle\rangle$ is infinite dimensional over $\langle\langle 1\rangle\rangle$.

In view of Corollary 25 and Theorem 26.1 condition 37.2 is definable over D. Thus Lemma 37 proves Theorem 38I.1.

Lemma 38. For t,x in D with t transcendental over $\langle\langle 1\rangle\rangle$ the following are equivalent:

1. $x = t^n$ for some integer n (possibly $n < 0$).
2. $x \in \langle\langle t\rangle\rangle$ and for any automorphisms h_2, h_3 of $\langle\langle t\rangle\rangle$ satisfying:

   $h_2(t) = t^2;\quad h_3(t) = t^3;$

   we have:

   $h_2(x) = x^2;\quad h_3(x) = x^3.$

Proof: Evidently $1 \Rightarrow 2$. The converse may be written more concretely as follows:

(*) If $r(t)$ is a rational function of t satisfying

$r(t^2) = (r(t))^2;\; r(t^3) = (r(t))^3;$

then $r(t) = t^n$ for some n.

We leave this as an exercise to the reader (Exercise 18).

Lemma 39. For t,x in D with t transcendental over $\langle\langle 1\rangle\rangle$ the following are equivalent:

1. $x = t^m$ for some nonnegative integer m.
2. $x = t^n$ for some integer n, and for any transcendental element y of D such that $y^t = y^2$, y^x is in $\langle\langle y\rangle\rangle$.

Proof: The point here is that if $y^t = y^2$, then $y^{t^n} = y^{2^n}$ for all n (positive, zero, or- suitably understood- negative).

Of course for y transcendental y^{2^n} is in $\langle\langle y\rangle\rangle$ iff $n \geq 0$.⊣

Remark. Applying Corollary 25 and Theorem 26.1, Lemmas 38-9 prove Theorem 36 II.1; and then Theorem 36 II. 2,4,5 are trivial. For example:

$x \oplus_t y = z$ iff $x,y \in N_t$ and $x \cdot y = z$.

The next lemma proves Theorem 36 II.3.

Lemma 40. For t transcendental over $\langle\langle 1\rangle\rangle$, x,y,z in D, the following are equivalent:

1. $x \otimes_t y = z$.
2. $x,y \in N_t$, and for any automorphism h of $\langle\langle t\rangle\rangle$, if $h(t) = x$ then $h(y) = z$.

Proof: Clear by inspection.

Thus we have proved Theorem 36.I.1, II. 1-5. Theorem 36 III.1 is obvious. We will dispose of 36 I.2, IV 1-2 before dealing with the main issue: Theorem 36 III.2.

Proof of Theorem 36 I.2, IV 1-2:

36 I.2: In view of the preceding lemmas we may define both the set N_t and the usual algebraic structure of arithmetic on N_t. In section 36 I.2 we seek a definition of the following predicate:

"The elements s^{t^n} commute with one another, and for each n the element s^{t^n} is transcendental over $\langle\langle s^{t^m} : m < n\rangle\rangle$."

This is almost trivially definable, in view of the foregoing; but the final clause requires a slight generalization of Lemma 37, which we leave without further ado to the reader.

36 IV.1: The statement "some element is transcendental over $\langle\langle 1\rangle\rangle$" is existential (Lemma 37, Corollary 25, Theorem 26.1) and hence true in D.

36 IV.2: If D is in DR^{∞}, we seek suitable parameters in D (compare the Convention in force since Lemma 37). Clearly D may be extended to a larger division ring D' in which there are suitable parameters (check). We may take D' also in DR^{∞}. Then since D is an elementary substructure of D', by 36 I.2 there are also suitable parameters in D.

Lemma 41. Let D be an $\aleph_1$-existentially saturated structure (Chapter 0 §6). Let s,t in D be such that t is transcendental over $\langle\langle 1\rangle\rangle$ and s,t are suitable. Then:

$^2N_{st} \simeq {}^2N$

(here it is understood that in $^2N_{st}$ subsets of N_t are coded by equivalence classes of elements of D- compare Definition 35.8).

<u>Proof</u>: Define $s_n = s^{(t^n)}$. Then in $^2N_{st}$ an element a of D codes the set $S_a = \{t^n: s_n{}^a = s_n\}$. Our claim is that every set S is coded by some element of D, in other words that for every $S \subseteq N_t$ there is an a such that:

$$s_n{}^a = s_n \quad \text{iff} \quad t^n \in S.$$

Since D is assumed $\aleph_1$-existentially saturated it suffices to verify that such an element can be found in an extension of D. This follows at once from Theorem 24 (existence of inner automorphisms) if we note that

$$\langle\langle s_n: n \geq 0 \rangle\rangle$$

is simply the (commutative) field generated over $\langle\langle 1 \rangle\rangle$ by the transcendence base $\{s_n: n \geq 0\}$.

Applying Lemma 41 we easily verify Theorem 36 III.2, completing the proof of Theorem 36. For any $D \in DR^\infty$ and suitable parameters s,t in D we consider the model $^2N_{st}(D)$. We may embed D in an $\aleph_1$-existentially saturated division ring D' in DR^∞, then $D < D'$ and

$$^2N_{st}(D') \simeq {}^2N.$$

One verifies easily that $^2N_{st}(D) \prec {}^2N_{st}(D')$, and hence

$$^2N_{st}(D) \equiv {}^2N, \text{ as claimed.}$$

<u>The proof of Theorem 36 is complete</u>.

Whenever one can prove a theorem resembling Theorem 36, a theorem of the following type follows immediately.

<u>Theorem 42</u>.

1. $DR^f \cap DR^\infty = \emptyset$
2. For every n $DR^n \neq DR^{n+1}$ (Chapter III §2).
3. There are $2^{\aleph_0}$ elementarily inequivalent existentially complete division rings of any desired characteristic.

We find it necessary here to invoke various recursion theoretic facts. We will give a correct but nonrigorous account of the relevant information, and refer to [25] for more detailed formulations.

Definition 43.

1. Let S,T be two theories. We say that S is reducible to T iff there is an effective procedure P for accomplishing the following:

Given a sentence X, find a sentence X* such that X is in S iff X* is in T.

2. Concerning coding: sentences of first order logic may be coded by numerals (as the computer programmer knows, any finitary object may be coded by numerals). In particular theories may be coded by sets of numerals. Hence first order statements about sets of numbers in ${}^2N$ may be construed as statements about theories (and vice versa).

Facts 44.

1. $Th({}^2N)$ is not reducible to $Th({}^1N)$.

2. The theory of ${}^2N$, coded by a set of numbers as in Definition 43.2, is not a definable element of ${}^2N$.

Proofs of such facts may be found in [39].

In the rest of this section we will make constant use of Definition 43.2 without explicit reference; in other words we will tend to ignore the distinction between sets of numbers and theories.

Proof of Theorem 42.1-2:

42.1. We will prove that for each p the theories $Th(DR_p^f)$, $Th(DR_p^\infty)$ are distinct. Since these theories are complete (Exercise 18) it will follow that $DR_p^f \cap DR_p^\infty = \emptyset$.

Suppose on the contrary that $Th(DR_p^f) = Th(DR_p^\infty)$. We claim that we then have the following situation, which will patently contradict Fact 44.1:

1. $Th({}^2N)$ is reducible to $Th(DR_p^\infty)$.

2. $Th(DR_p^\infty) = Th(DR_p^f)$.

3. $Th(DR_p^f)$ is reducible to $Th({}^1N)$.

The first point is merely a summary of Theorem 36, and the second point has been assumed. We may therefore confine ourselves to the verification of point 3.

If we take into account the material of Chapter III §3 it will become clear (upon reflection) that $Th(DR_p^f)$ is reducible to $Th({}^1N)$. Most of the details will be left to the reader, but the following sketch should suffice.

Let X be a sentence in the language of division rings. We seek a sentence X^* such that

(*) $X \in Th(DR_p^f)$ iff $X^* \in Th({}^1N)$.

Assuming for simplicity that X is in prenex normal form,

$X = "\forall x_1 \exists x_2 \ldots"$

we see that the following are equivalent:

i. $X \in Th(DR_p^f)$.

ii. DR_p^f forces $\neg\neg X$.

iii. For any condition m there is a condition n extending m such that n forces X.

iv. For any condition m there is a condition n extending m such that for any condition n_1 extending n there is a condition n_2 extending n_1 and a constant c_2 such that ....

In clause iv above, we imagine the definition of forcing written out in glorious detail (with conditions coded by numbers). Thus iv may be construed as a number theoretic statement X^*, and (*) holds.

<u>42.2</u>. If $DR^n = DR^{n+1}$ then $DR^n = DR^\infty$. On the other hand we will show that $Th(DR^n) \neq Th(DR^\infty)$. This follows from the following two observations:

1. The theory of DR^∞ is not a definable set of numbers in ${}^2N$.

2. The theory of DR^n is a definable set in ${}^2N$.

The first observation follows from Fact 44.2, taking into account the reducibility of $Th(^2N)$ to $Th(DR^\infty)$. The argument is as follows. For each sentence X in the language of ^2N let X* be the corresponding sentence in the language of division rings. Then

$$Th(^2N) = \{X: X^* \text{ is in } Th(DR^\infty)\}.$$

Thus if $Th(DR^\infty)$ were definable in ^2N, $Th(^2N)$ would also be definable in ^2N, in contradiction to Fact 44.2.

The second observation involves a fairly straightforward verification in the manner of the proof of 42.1. More explicitly, the following are equivalent:

1. $X \in Th(DR^n)$.
2. For every countable model S in DR^n, X is true in S.
3. For every countable model S, either:

   i. X is true in S

   or ii. S is not in DR^n - i.e. there is a DR^{n-1}-persistent sentence X and a countable model T such that $S \subseteq T$, T is in DR^{n-1}, X holds in T, $\neg X$ holds in S.

We claim that clause 3 above can be expressed by a sentence X* of second order number theory. To avoid tedium we confine ourselves to the following remarks:

In the first place, <u>countable</u> models S may be coded by sets of integers. Secondly, we may assume inductively that we already know how to interpret clause 3ii in second order number theory (since it involves DR^{n-1} rather than DR^n).

This concludes the proof of 42.2.

A far more careful version of the above is found in [25].

We come now to the proof of Theorem 42.3. We will continue our policy of giving vague accounts of recursion theoretic notions, relying on [25] for rigor. In particular we now need the notions of Σ_1^1 and Δ_1^1 sets.

Definition 45. Since there are two sorts of objects in ^2N, the quantifiers $\forall x$ and $\exists x$ may refer to both numbers and sets of numbers. It is convenient at this point to change our language slightly by making a distinction between two sorts of variables:

number variables x,y,z ranging over numbers

set variables X,Y,Z ranging over sets of numbers.

We will then say that a definable subset S of ^2N is:

1. Σ_1^1 iff S is definable in ^2N by a formula $F(x)$ involving at most a single existential quantification over sets of numbers and arbitrary quantification over numbers (no universal set quantifier is permitted).
2. Δ_1^1 iff both S and its complement are Σ_1^1.

We will derive Theorem 42.3 from the following recursion theoretic fact.

Fact 46. Let S be a Σ_1^1 set of sets of integers, and suppose S contains a member T which is not a Δ_1^1 set. Then S has cardinality $2^{\aleph_0}$.

(A proof of this theorem may be found in [42]. It involves a careful variant of the usual splitting argument; compare the proof of Theorem 27.)

Proof of Theorem 42.3.

We consider the set S of all complete theories of existentially complete division rings of fixed characteristic. We will prove that S is a Σ_1^1 set with a non-Δ_1^1 member, and invoke Fact 45.

S is Σ_1^1: A theory T is in S iff:

(*) there is a complete theory T' involving constants $\{c_i : i<\infty\}$ such that T' is the theory of a countable existentially complete division ring $D = \langle \{c_i\}; +,\cdot \rangle$ and T is the set of sentences of T' which do not involve the constants c_i.

We claim that (*) provides a Σ_1^1 definition of S. Certainly

(*) involves an existential set quantifier ("there is a theory T' "); we must see that the rest of (*) is harmless (i.e. can be formalized without using set quantifiers). After cursory inspection we see that the key clause is:

(**) T' is the complete theory of an existentially complete division ring $D = \langle c_i ; +, \cdot \rangle$.

For (**) to hold, T' must be complete and consistent. On the other hand, if T' is complete and consistent then the only possible candidate for D is the division ring whose addition and multiplication tables are explicitly described by T'. Call this division ring D(T'). Then (**) amounts to the following:

1. T' is a complete consistent extension of DR.
2. T' is the theory of D(T').
3. D(T') is existentially complete.

Conditions 1-3 may easily be rephrased in terms of formal properties of the theory T' (condition 1 is already in that form); and these properties may easily be expressed in number theoretic terms without use of set quantifiers. To see that this is the case it is necessary to reformulate conditions 2,3 slightly. The details are indicated in Exercise 19.

<u>S contains an element T which is not Δ^1_1</u>:

Let T be the theory of an infinitely generic division ring. We observed above that T is not definable in ^2N (proof of 42.2). In particular neither T nor its complement could possibly be Σ^1_1-definable. Of course T is in S.

Thus by Fact 46 S has cardinality $2^{\aleph_0}$, and the result follows. Again, this theorem is proved more carefully in |25|.

§4. Existentially Complete Groups.

The material of §§2-3 applies with minor alterations to existentially complete groups. The proof of the amalgamation property for groups is comparatively straightforward [45]. In the analog of Theorem 36 models of arithmetic are constructed using powers of an element x of infinite order rather than powers of a transcendental element. There are no surprises.

This section is devoted to a theorem concerning finitely generated subgroups of existentially complete groups. It is assumed that the reader is familiar with presentations of groups via generators and relations. We recall that a set S is said to be recursive iff there is an effective procedure for deciding whether or not any particular object x is a memeber of S (cf. [39]).

Definition 47. Let G be a finitely generated group.

1. G is recursively (respectively, finitely) presented iff G has at least one presentation with a recursive (respectively, finite) set of relations.

2. G has solvable word problem iff the set of all relations holding in G is recursive.

(Notice that the content of these definitions is independent of the choice of a specific set of generators for G.)

The main theorem of this section is the following.

Theorem 48. Let G be a finitely presented group. Then the following are equivalent:

1. G is embeddable in every existentially complete group.

2. G has solvable word problem.

We will make use of one extremely powerful auxiliary result.

Fact 49. Let G be a recursively presented finitely generated group. Then G is embeddable in a finitely presented group.

A proof of this theorem may be found in the appendix to [50].

Proof of Theorem 48:

If the finitely generated group G is embedded in a group H and I is a subset of H, say that G is <u>rigidly embedded</u> in H relative to I iff for every homomorphism h with domain H,

$\ker(h) \cap I = \emptyset \Rightarrow h$ is injective on G.

Consider the following condition on G:

3. G can be rigidly embedded in a finitely presented group H relative to a finite subset I of H.

We claim that conditions 1-3 are equivalent. The simplest and most illuminating of the implications is:

<u>3 => 1</u>: Let G be rigidly embedded in the finitely presented group H relative to the finite subset I of H. Let H have generators $h_1,\dots,h_k$ and relations $r_1,\dots,r_l$. Let $I = \{i_1,\dots,i_m\}$. Consider the existential statement:

(e) $\exists h_1\dots h_k \ (r_1,\dots,r_l = 1$ and $i_1,\dots,i_m \neq 1)$.

This is true in H; hence it is true in any existentially complete group E.

Let E be any existentially complete group. Since (e) says that there is a homomorphism $h: H \to E$ whose kernel is disjoint from I, the definition of rigidity implies that h embeds G in E. Thus 3 => 1. We will now prove the converse.

<u>1=>3</u>: Assume that $G = \langle g_1,\dots,g_k\rangle$ is embeddable in every existentially complete group. Nevertheless, let us try to construct an existentially complete group E containing no copy of G. We will see that the failure of this enterprise will produce a rigid embedding of G into a finitely generated group H relative to a finite set I.

We propose to construct the group E in infinitely many stages by successively imposing relations on the free group E_0 on infinitely many generators $x_1, x_2, \dots$. At stage n we will have two finite lists R_n (relations) and I_n (irrelations) of words, and a finitely presented group E_n, subject to the following general conditions:

1. E_n is the quotient of E_0 by the relations R_n.
2. Each word i in I_n is unequal to the identity in E_n (i.e. the relations R_n are consistent with the irrelations I_n).
3. $R_{n-1} \subseteq R_n$, $I_{n-1} \subseteq I_n$.

Having completed the construction we will set $R = \cup R_n$, $I = \cup I_n$, and let E be the quotient of E_0 by the relations R. In E the irrelations $i \neq 1$ (i in I) will also hold. It will be convenient to adopt the following abbreviated terminology: we say that a group H satisfies the relations R and the irrelations I just in case H satisfies the relations $r = 1$ $(r \in R)$ and the irrelations $i \neq 1$ $(i \in I)$.

Our objective is twofold:

I. Make E existentially complete.

II. Ensure that G is not embeddable in E.

More concretely: first make a list of all existential sentences e in the language of E_0, and list all k-tuples $w_1,\dots,w_k$ of words in E_0 (recall $G = \langle g_1,\dots,g_k\rangle$). At stage $n+1$ in our construction we will try to ensure that:

I'. If there is a group H satisfying the relations R_n and the irrelations I_n in which the nth existential sentence e is true, then e is true in <u>every</u> group satisfying the relations R_{n+1} and the irrelations I_{n+1}.

II'. Let $w_1,\dots,w_k$ be the nth k-tuple of words in E_0. Then for any group H satisfying the relations R_{n+1} and irrelations I_{n+1}, the function $g_i \longrightarrow w_i$ cannot be extended to an isomorphism from G to $\langle w_1,\dots,w_k\rangle_H$.

Certainly if we can systematically accomplish I',II', then we will have I,II. There is in fact little difficulty in treating requirement I', since an existential sentence e merely asserts the existence of a handful of elements satisfying various relations and irrelations, and since R_n and I_n involve only finitely many generators x_i of E_o. Since there are no complications, we leave the verification to the reader.

Life becomes more interesting when we try to ensure II'. Let W_n be the group generated by $w_1, \ldots, w_k$ in E_n. There are two cases:

i. The map $g_i \longrightarrow w_i$ cannot be extended to an isomorphism of G with W_n.

ii. The map $g_i \longrightarrow w_i$ can be extended to an isomorphism of G with W_n.

The first case is trivial: there must be a relation r or irrelation i holding in W_n but not in G. Adjoin r or i, to the appropriate list, and II' is ensured.

In the second case, if W_n is <u>not</u> rigidly contained in E_n relative to I_n, we may impose an <u>additional</u> relation on W_n consistent with I_n (see the definition of rigidity). On the other hand if W_n <u>is</u> rigidly contained in E_n with respect to I_n, then the isomorphism of G with W_n provides a rigid embedding of G into E_n with respect to I_n, and we may easily truncate E_n to a finitely generated subgroup with the same property. Thus we have the desired rigid embedding of G into a finitely presented group (of course in this case the construction of E breaks down).

Since we know that the preceding construction must fail sooner or later, it follows that at some point we will construct a rigid embedding of G into a (finitely generated subgroup of a) group E_n with respect to a finite set I_n. Since E_n is finitely presented we have proved that 1 => 3.

Thus 1 <=> 3.

<u>3 => 2</u>: Suppose that G is rigidly embedded in the finitely presented group H relative to the finite set I. Let R be the finite set of relations in a finite presentation of H. Then all relations <u>and irrelations</u> holding in G are consequences of the

finite set of relations R and irrelations I. It is not difficult to systematically enumerate all relations and irrelations entailed by R,I. It follows readily that G has solvable word problem.

2 => 3. Assume $G = \langle g_1,\dots,g_k\rangle$ has solvable word problem. We use the exponential notation $a^b = b^{-1}ab$. Make an effective list of all words w in the generators $g_1,\dots,g_k$ such that $w \neq 1$ in G, and let G' be the group generated over G by elements a,b,t modulo the relations:

$$(*) \qquad t = w_i^{a^{b^{2i}}} w_i^{a^{b^{2i+1}}} .$$

This makes a certain amount of sense since in the free group $\langle a,b,t\rangle$ the elements $c_i = a^{b^i}$ generate a free group $\langle c_i\rangle$, and we are just imposing the relations

$$t = w_i^{c_{2i}} w_i^{c_{2i+1}} .$$

It is not difficult to see that G is embedded in G' (we leave the combinatorics to the reader as Exercise 20). Furthermore G' is recursively presented, hence embeddable by Fact 49 in a finitely presented group H. Finally notice that G is rigidly embedded in G' (hence in H) relative to t ; for evidently (*) implies:

$$t \neq 1 \Rightarrow w_i \neq 1.$$

Thus 2 <=> 3.

§5. Notes.

The amalgamation problem for division rings was studied by Cohn in a number of papers going back to [13], and successfully resolved in [15]. The article [17] provides a readable brief account of some of this work, and a great deal of material is found in the book [16].

Our account of the algebraic and model theoretic properties of existentially complete division rings follows Wheeler [25]. There is a great deal of overlap between the work of Wheeler and Macintyre (cf. [34, 54]). The proof given here for Theorem 24 was suggested by Macintyre.

Theorem 48 is due to Neumann (2 => 1) and Macintyre (1 => 2). Theorem 48 has been variously extended (e.g. in [10]). The powerful Fact 49 is better known as Higman's Embedding Theorem [50].

Exercises

§1.

We assume that $A \subseteq B_1, B_2$ are division rings.

1. Let R be a ring admitting a nonzero homomorphism into a division ring. Show that any two free modules over R of different dimensions are nonisomorphic. (Compare Exercise 7.) In particular this applies to commutative rings with identity.

2. Viewing B_1, B_2 as right A-vector spaces, fix decompositions

   $$B_1 = A \oplus B_1'$$

   $$B_2 = A \oplus B_2'.$$

   For any sequence $I = (i_1, \dots, i_n)$ of alternate 1's and 2's form

   $$B_I = B_{i_1}' \otimes \dots \otimes B_{i_n}' \quad \text{(tensor products are over } A\text{)};$$

   for $I = ()$ let $B_I = A$.

   Write $R = \oplus\sum B_I$. Show that R admits a multiplication such that R becomes the universal product of B_1, B_2 over A.

   In more detail, define

   $$^{i}R = \oplus\sum_{\substack{I = (i_1 \dots i_n) \\ i_1 = i}} B_I$$

   and notice the natural isomorphism $R \simeq B_i \otimes {}^{i}R$. This gives R a left (B_1, B_2)-bimodule structure compatible with the right A-module structure on R. In particular we have a natural homomorphism φ_0: $A \oplus B_{(1)} \oplus B_{(2)} \longrightarrow \mathrm{End}(R)$, the endomorphism ring of the right A-module R. Show that this map extends to a homomorphism $\varphi: R \longrightarrow \mathrm{End}(R)$, and that multiplication may be defined by $r_1 \cdot r_2 = \varphi(r_1) r_2$.

3. Prove that a fir has no zero divisors, and a commutative fir is a principal ideal domain.

4. Prove the Interaction Lemma. Specifically, let a,b be elements of the ring $R = B_1 *_A B_2$, where a is right pure of type 1 and height m and b is left pure of type 1 and height n. Assume:

   $h(ab) < m+n-1$.

   Prove that a is right reducible (and b is left reducible).

   Some details: fix bases $\{e_i\}, \{f_j\}$ of B_1, B_2 ($e_o = f_o = 1$), and construct the natural bases $\{u_j\}$, $\{u_j, v_k\}$ of H_{m-2}, H^2_{m-1} (elements of height m-2 or of height m-1 and right type 2 respectively). Write $a = \sum a_i e_i$ with $a_i \in H^2_{m-1}$,

   $a_i = \sum v_k \alpha_{ki} + a_i^*$, $\alpha_{ki} \in A$, $a_i^* \in H_{m-2}$.

   Setting $x_k = \sum \alpha_{ki} e_i$ compute that

   I. $a = \sum v_k x_k + a^*$ $\quad v_k \in H^2_{m-1}$, $x_k \in B_1$, $a^* \in H_{m-2}$.

   II. $x_k b \in H_{n-1}$.

   Analyze b in similar fashion, obtaining

   III. $b = \sum y_l w_l + b^*$

   IV. $x_k y_l \in A$.

   Show that $a y_1 \in H_{m-1}$, so that a is right reducible.

5. Show that $B_1 *_A B_2$ has no zero divisors.

6. $R = B_1 *_A B_2$. Use the Weak Algorithm (Theorem 14) to prove that every right (or left) ideal I of R is free. The following notion will be useful: a set $X \subset R$ is <u>right</u> <u>R-dependent</u> iff X contains an n-tuple $\bar{a}$ such that for some n-tuple $\bar{b}$ we have:

   $h(a_i b_i) \equiv N$ but $h(a.b) < N$. Otherwise we say X is right R-independent.

   (Choose a very large right independent set of right irreducible elements of I; with a little care one will get a basis for I.)

7. Let M,N be two free modules of different dimensions over a ring R. Assume M is infinite dimensional. Show that M,N are non-isomorphic.

8. $R = B_1 *_A B_2$. Use the Weak Algorithm (Theorem 14) to show that R is a fir. By exercises 6,7 it suffices to prove that

$$R^m \not\cong R^n$$

for $m \neq n$ (m,n are finite integers). The main steps are:

I. The statement to be proved may be reformulated as follows: there are no matrices A (mxn), B (nxm) over R whose products are identity matrices-

$AB = I_m$, $BA = I_n$.

II. The weak algorithm has the following crucial consequence:

(*) if $\bar{a}, \bar{b} \in R^n$, $\bar{a} \neq 0$, then there is an invertible nxn matrix A such that $A\bar{b}$ has a zero entry.

In fact, if an invertible matrix A^{-1} is chosen so that the n-tuple:

$(h'(a_1'),\dots,h'(a_n'))$ $\quad(\bar{a}' = \bar{a}A^{-1}$, h' is modified height)

is minimal in the natural partial ordering of $(1/2\,\underset{\sim}{N})^n$, then some coordinate of $A\bar{b}$ will be zero.

III. Apply (*) to prove I. (Slight hint: $AB = AX^{-1}XB$ for all invertible X.)

9. In order to construct a division ring from scratch it is normally necessary to construct two binary operations +,· satisfying numerous axioms. However it turns out that the structure of a division ring D is already determined by:

multiplication on the group G of nonzero elements and the operation $x \longrightarrow 1-x$ defined on $G-\{1\}$.

Prove the above in the following formulation:

Assume G is a group and $\theta: G-\{1\} \longrightarrow G-\{1\}$ is a map satisfying:

i. θ^2 = identity.

ii. $\theta(x^y) = (\theta x)^y$.

iii. $\theta(xy^{-1}) = \theta((\theta x)(\theta y)^{-1})\theta(y^{-1})$.

iv. $(\theta(x^{-1}))x(\theta x)^{-1} = e$ is independent of x.

Then G is the multiplicative group of a unique division ring D such that $\theta x = 1-x$ and $e = -1$.

10. Let

$$0 \longrightarrow K \longrightarrow F \longrightarrow M \longrightarrow 0$$

$$0 \longrightarrow K' \longrightarrow F' \longrightarrow M \longrightarrow 0$$

be two presentations of a module M with F, F' free. Prove that $K\oplus F' \simeq K'\oplus F$. (Show in fact that both of these modules are isomorphic to the submodule of $F\oplus F'$ consisting of pairs (f,f') where f,f' represent the same element of M.)

11. Let T be a full $n\times n$ matrix over a fir R. View T as an endomorphism of R^n. Let $\ker T = K$, range $T = F$. Show that K,F are free and $R^n \simeq K\oplus F$. Conclude that $K = (0)$.

In Exercises 12-13 we use the notation of Theorem 23.

12. Assume T is full. Prove that M is a torsion module. (Hint: for $N \subsetneq M$ consider the diagram

$$\begin{array}{ccccc} R^n & \xrightarrow{T} & R^n & \xrightarrow{\pi} & M \\ \| & & \uparrow & & \uparrow \\ R^n & \xrightarrow{T} & \pi^{-1}(N) & \xrightarrow{\pi} & N. \end{array}$$

Show that $\pi^{-1}(N)$ is free and hence has rank at least n, since T is full. Deduce $rk(N) \geq 0$.)

13. Assume M is a torsion module. Show T is full. (If $T = AB$ with A $n\times m$, B $m\times n$, $m<n$ then consider the diagram

$$\begin{array}{ccccccc} & & R^n & \xrightarrow{B} & R^m/\ker A & & \\ & & T\downarrow & & \downarrow A & & \\ & & R^n & = & R^n & & \\ & & \downarrow & & \downarrow & & \\ 0 \longrightarrow K \longrightarrow & & M & \longrightarrow & M' & \longrightarrow 0 & \end{array}$$

Compute $rk(M') > 0$, a contradiction.)

§2.

14. Let T be a universal theory with the amalgamation property.

Let D be a model of T, $A,B \subseteq D$ submodels of D. Let $\varphi: A \simeq B$. Show that D has an extension D' such that φ can be extended to an automorphism of D'.

15. Let R be a ring without zero divisors. R is called a <u>right Ore domain</u> iff any two elements a,b in R have a common nonzero right multiple (i.e. there exist nonzero elements c,d in R such that $ad = bc$).
Show that any right Ore domain can be embedded in a division ring. In fact with minor variations the usual construction involving formal symbols a/b works as in the commutative case.

16. Let $D[t;h]$ be a skew polynomial ring as in the proof of Theorem 24. Show that $D[t;h]$ is right noetherian (the usual proof applies). Conclude as follows that $D[t;h]$ is a right Ore domain: if a,b are in $D[t;h]$ prove that the elements $\{a^n b\}$ are (right) linearly dependent over R and deduce that a,b have a common right multiple.

17. Verify Example 30.

§3.

18. Show that DR_p^f, DR_p^∞ are complete theories.

19. Let T' be a complete consistent extension of a theory T (in §3 $T = DR$ or DR_p). Assume T' involves the constant symbols $\{c_i : i<\infty\}$ and no others. Let M be the model described by the quantifier free sentences in T' (M consists of equivalence classes of constant symbols modulo the equivalence relation: $c \sim c'$ iff "$c = c'$" $\in T'$; a relation Rcc' holds in M iff "Rcc'" $\in T'$; etc.). Then:

(Henkin) T' is the theory of M iff for every sentence "$\exists x\, F(x)$" $\in T'$ there is a constant c such that "$F(c)$" $\in T'$.

(Hirschfeld) If T' is the theory of M, then M is existentially complete as a model of T iff for every existential sentence e <u>not</u> in T' there is an existential sentence e' in T' such that T proves "e' => ¬e."

V. Existentially Complete Modules

Introduction.

So far we have become acquainted primarily with two sorts of theories T:

1. Theories with model companions (Fields and Ordered Fields, Chapter I §2; Formally p-adic Fields, Chapter II §8; Semiprime Commutative Rings, Chapter III §5).

2. The diametrically opposed case: T has no model companion, $T^f \cap T^\infty = \emptyset$, the hierarchy $\{T^n\}$ is nondegenerate, and $Th(T^\infty)$ is even nonanalytic.

In this chapter we study an intermediate situation. Let R be a ring and let $T = M_R$ be the theory of R-modules (if the ring R is uncountable then this involves the use of an uncountable language). Consider the following condition on R:

(*) M_R has a model companion.

This is equivalent to an algebraic condition on R, given in §4. From the present point of view the contrary case is more interesting. Even when M_R has no model companion, we see in §5 that $\underset{\sim}{E}_{M_R} = M_R^f = M_R^\infty$; in this sense M_R "almost" has a model companion.

(We will take "module" to mean "right module"; similarly "ideal" will mean "right ideal".)

§1. Z-modules.

Let AG be the theory of abelian groups. In its natural formulation AG is not quite identical with the theory of Z-modules, but the differences are inessential. As preparation for §§2-3 we are going to make a brief study of $\underset{\sim}{E}_{AG}$. AG will be studied in more depth from another point of view in Chapter VI.

Let AG* be the following theory of abelian groups A:

1. AG.
2. "A is divisible" (For every a in A and every integer n, n divides a in A.)
3. "There are infinitely many elements of each finite order."

Theorem 1. AG* is the model companion of AG.

Before beginning the proof we recall a few facts about divisible abelian groups. The canonical examples of such groups are the additive group Q of the rationals and the Prüfer groups Z/p^{∞} of all complex p^nth roots of 1 (n varies), for any prime p. (Cf. Chapter VI §1.)

The following property of divisible groups is extremely useful:

Proposition 2. A is divisible iff A is a direct summand of any abelian group containing A.

Proof: Exercises 1,2.

Using Proposition 2 and some argument, one gets the following structure for divisible groups (Proposition VI.1.7):

(⊕) $A = \oplus\Sigma\,(Z/p^{\infty})^{(\gamma_p)} \oplus Q^{(\delta)}$,

where γ_p, δ are invariants of the group A.

Proof of Theorem 1:

Our claim is that $\underset{\sim}{E}_{AG} = \mathrm{Mod}(AG^*)$. Since every model of AG evidently extends to a model of AG* it should be clear that:

$\underset{\sim}{E}_{AG} \subseteq \mathrm{Mod}(AG^*)$.

Conversely if A satisfies AG* we must show that A is existentially complete.

Assume therefore that $A \subseteq B$ and B satisfies an existential sentence $e(a_1,\ldots,a_n)$ with $a_1,\ldots,a_n$ in A. We claim A satisfies $e(a_1,\ldots,a_n)$. More precisely let

$e(\bar{a}) = \exists\bar{x}\; e_0(\bar{a},\bar{x})$

where e_o contains no quantifiers, and choose $\bar{b}$ in B satisfying $e_o(\bar{a},\bar{b})$. We will complete the following diagram:

$$\text{(D)}\qquad \begin{array}{ccc} \langle\bar{a},\bar{b}\rangle & \overset{f}{\dashrightarrow} & A \\ \nwarrow & & \nearrow \\ & \langle\bar{a}\rangle & \end{array}\ ;$$

then $e_o(\bar{a},f(\bar{b}))$ holds in A, hence $e(\bar{a})$ holds in A.

To complete diagram (D) use Proposition 2 to write $B = A \oplus C$, and then apply ($\oplus$) together with the axioms of AG* to write:

$$A = \oplus\Sigma\ (Z/p^{\infty})^{(\gamma_p)} \oplus Q^{(\delta)} \tag{$\oplus\oplus$}$$

with $\gamma_p \geq \aleph_o$.

Now write $b = b' + b''$ with b' in A and b'' in C. Since $\bar{a}$, b' contain only finitely many nonzero companents among the factors appearing in ($\oplus\oplus$), it is evident that diagram (D) may be completed by holding $\bar{a}$, b' fixed and moving b'' inside A. $\dashv$

As we shall see, this argument generalizes to modules over any ring R for which the analog of divisibility is well behaved. One way to generalize the notion of divisibility is to use Proposition 2 as a definition. However we see from the definition of AG* that we need a notion of divisibility which can be axiomatized by first order statements.

§2. R-<u>modules</u>.

<u>Convention</u>. R is a fixed ring. M_R is the theory of R-modules.

In formalizing M_R we use the function + together with the multiplication functions μ_r ($\mu_r(x) = xr$) for each r in R. In particular the following axioms are included in M_R:

(1_{rs}) $\qquad \forall x\, \mu_{r+s}(x) = \mu_r(x) + \mu_s(x)$

(2_{rs}) $\quad \forall x\, \mu_{rs}(x) = \mu_s(\mu_r(x)).$

The theory M_R is universal. We recall (Chapter III §1) that universal theories having the amalgamation property and a model companion admit elimination of quantifiers.

Theorem 3. M_R has the amalgamation property.

Proof: If

$$M \begin{array}{l} \nearrow M_1 \\ \searrow M_2 \end{array}$$

is a diagram of R-modules and embeddings, we define $M_1 \oplus_M M_2$ as the quotient of $M_1 \oplus M_2$ by the submodule $M' = \{(m,-m)\colon m \text{ in } M\}$. Then the canonical injections $j_i\colon M_i \longrightarrow M_1 \oplus M_2$ induce embeddings $j_i^*\colon M_i \longrightarrow M_1 \oplus_M M_2$ such that:

$$M \begin{array}{l} \nearrow M_1 \xrightarrow{j_1^*} \\ \searrow M_2 \xrightarrow[j_2^*]{} \end{array} M_1 \oplus_M M_2.$$

We omit the verifications.

Definition 4. $M_1 \oplus_M M_2 = (M_1 \oplus M_2)/\{(m,-m)\colon m \in M\}$. This is called the direct sum of M_1, M_2 with amalgamated submodule M (it is also called a pushout, but not by us).

Notice that $M_1 \oplus_M M_2$ is defined even if the maps

$f\colon M \longrightarrow M_1$, $g\colon M \longrightarrow M_2$

are not 1-1 (interpreting "$(m,-m)$" as "$(fm,-gm)$"). This remark will be of use to us in Theorem 5 and occasionally thereafter.

Let us now consider how the material of §1 can be generalized to M_R. We seek a model companion M_R^* of M_R (when it exists) of roughly the following form:

1. M_R.
2. M is "divisible" (see below).
3. M contains a variety of elements annihilated by various

ideals of R.

The second point represents the heart of the matter. For general modules the analog of divisibility is <u>injectivity</u>, defined below (Definition 6). The following theorem deals with various possible generalizations of the notion of divisibility.

<u>Theorem 5</u>. Let M be an R-module. Then the following are equivalent:

1. M is a direct summand of any R-module containing M.

2. Any diagram of the following form can be completed:

(D1) $$\begin{array}{ccc} M_1 & \xrightarrow{1-1} & M_2 \\ h\downarrow & \swarrow & \\ M & & \end{array}$$ (h is not assumed to be 1-1)

3. Any diagram of the following form can be completed:

(D2) $$\begin{array}{ccc} I & \xrightarrow{1-1} & R \\ h\downarrow & \swarrow & \\ M & & \end{array}$$ (I is an ideal of R; h is not assumed to be 1-1.)

4. Any system of equations $\{xr = m_r$: r varies over a set $S \subseteq R$, m_r are in $M\}$ which has a solution in an extension of M is already solvable in M.

<u>Definition 6</u>. A module M is <u>injective</u> iff M satisfies 5.1-4.

<u>Proof of Theorem 5</u>:

<u>1 => 2</u>: Let (D1) be as shown, and take N to be the direct sum of M, M_2, amalgamating M_1. One checks easily that in

$$\begin{array}{ccc} M_1 & \xrightarrow{1-1} & M_2 \\ \downarrow & & \downarrow \\ M & \longrightarrow & N \end{array}$$

the map $M \longrightarrow N$ is 1-1. Thus M is a direct summand of N and there is a projection map $\pi: N \longrightarrow M$. Then complete (D1) using the map

$M_2 \longrightarrow N \xrightarrow{\pi} M$.

2 => 3: Instantaneous.

3 => 4: We assume that N is an R-module extending M and that N contains a solution n of:

$nr = m_r \quad (r \in S \subseteq R)$.

Define a map $f: R \longrightarrow N$ by $f(r) = nr$, and let I be the ideal $f^{-1}[M]$. Then:

$$\begin{array}{ccc} I & \subseteq & R \\ \downarrow & \swarrow g & \\ M & & \end{array}$$

can be completed.

Let $m = g(1)$. For r in S we compute

$mr = g(1)r = g(r) = f(r) = m_r$, as desired.

4 => 2: Exercise 2. (This is an imppertant argumeht.)

2 => 1: If $M \subseteq M_1$, complete the diagram

$$\begin{array}{ccc} M & \xrightarrow{1-1} & M_1 \\ = \downarrow & \swarrow \pi & \\ M & & \end{array} \quad ;$$

then $M_1 = M \oplus \ker \pi$.

Notice that the fourth condition is related to existential completeness in an infinitary language, or in ordinary first order logic if we restrict ourselves to finite sets of equations (or finitely generated ideals in 5.3).

For any cardinal κ (and notably for $\kappa = \aleph_0$), a module M is called κ-injective iff M satisfies 5.4 for sets of equations of cardinality less than κ (or 5.3 for ideals generated by fewer than κ elements). It develops that M_R has a model companion iff the property "M is $\aleph_0$-injective" is first order definable, and that this is the case iff R satisfies a reasonably well known condition weaker than the noetherian condition, called coherence. For our purposes, we may define:

Definition 7. R is coherent iff the class of $\aleph_0$-injective modules is first order axiomatizable.

We will come to the algebraic characterization of coherent rings in §4. Our immediate concern is the

<u>Theorem 8</u>. M_R has a model companion iff R is coherent.

We prove the necessity of the coherence condition here, and devote the next section to the converse.

<u>Proof of Theorem 8 (Necessity)</u>:

Assume M_R has a model companion M_R^*. Recall that in this case M_R admits elimination of quantifiers (Theorem 3 and preceding remark). Consider any finite set of equations:

(E) $xr = m_r$ $r \in F$ (a finite subset of R).

Then there is a quantifier free formula $w(\bar{m}_r)$ such that M_R^* proves:

(E*) $\forall \bar{m}_r \left[(\exists x \bigwedge_r xr = m_r) \Leftrightarrow w(\bar{m}_r) \right]$.

It follows easily that the $\aleph_0$-injective R-modules are axiomatized by $M_R \cup \{(E^*)$: E varies over finite sets of equations$\}$. ⊣

§3. <u>Existence of Model Companions</u>.

Our candidate for the model companion M_R^* of M_R, when R is coherent, will be the following:

1. "M is an $\aleph_0$-injective R-module"

2. "M contains a large number of elements annihilated by various ideals of R"

It is time to devote some attention to the second point.

<u>Definition 9</u>.

1. The sequence $r_1,\dots,r_k,s_1,\dots,s_l$ of elements of R is <u>relevant</u> iff for all j $s_j \notin \sum_i r_i R$.

2. For relevant $\bar{r},\bar{s}$ let

$w_{\bar{r},\bar{s}}(x) = \bigwedge_i xr_i = 0 \;\&\; \bigwedge_j xs_j \neq 0.$

3. For relevant $\bar{r},\bar{s}$ and $n \geq 1$ define:

$\mathrm{Ann}^{n}_{\bar{r},\bar{s}}$ = " $\exists x_1 \ldots \exists x_n \, (\bigwedge_i w_{\bar{r},\bar{s}}(x_i) \,\&\, \bigwedge_{i\neq j} \bigwedge_{p,q} x_i s_p \neq x_j s_q)$."

Lemma 10. Let k be a cardinal larger than the cardinality of R. Let M be a k^+-saturated R-module. Then the following are equivalent:

1. For every ideal I of R M contains at least k disjoint submodules isomorphic with the module R/I.
2. For all relevant $\bar{r},\bar{s}$ and $n \geq 1$ M satisfies: $\mathrm{Ann}^{n}_{\bar{r},\bar{s}}$.

Proof:

We leave this as Exercise 7.

The point is that the following conditions are equivalent:

i. $\langle x \rangle \cong R/I$

ii. the annihilator of x is I.

Definition 11.

An R-module M will be called fat iff M satisfies all of the conditions:

$\mathrm{Ann}^{n}_{\bar{r},\bar{s}}$

for $\bar{r},\bar{s}$ relevant and $n \geq 1$.

In order to complete the proof of Theorem 8 we will need some information concerning the existence of embeddings of various modules into injective modules. We defer the proof of the following theorem to the end of this section.

Theorem 12. Let M be an R-module. Then M is embeddable in an R-module $\bar{M}$ satisfying:

1. $\bar{M}$ is an injective extension of M.
2. For any injective extension N of M the following diagram can be completed:

$$\begin{array}{ccc} \bar{M} & \dashrightarrow & N \\ & \nwarrow \quad \nearrow & \\ & M & \end{array} .$$

Such a module $\bar{M}$ is called an <u>injective hull</u> of M.

The crux of the proof of Theorem 8 will be a characterization of sufficiently saturated existentially complete R-modules. How saturated is "sufficiently" saturated?

<u>Definition 13</u>. Let $\kappa(R)$ be the smallest cardinal greater than $\text{card}(R) + \aleph_0$ such that:

every finitely generated R-module M has an injective hull of cardinality $< \kappa(R)$.

Modules which are $\kappa(R)$-saturated are "sufficiently" saturated!

<u>Theorem 14</u>. Let M be a $\kappa(R)$-saturated R-module. Then the following are equivalent:

1. M is existentially complete.
2. M is $\aleph_0$-injective and M is fat (Definition 11).

<u>Proof</u>: <u>1 => 2</u> is evident and makes no use of the saturation hypothesis.

<u>2 => 1</u>: Since M is at least $\text{card}(R)^+$-saturated and $\aleph_0$-injective it follows that M is injective (using Theorem 5.4). Since M is fat Lemma 10 tells us that for each $I \subseteq R$, M contains at least $\kappa(R)$ disjoint copies of R/I.

Suppose now that $\bar{m} \subseteq M$, $e(\bar{m})$ is an existential formula, and $e(\bar{m})$ is true in an extension N of M. We will show that $e(\bar{m})$ holds already in M. More precisely let

$e(\bar{m}) = \exists \bar{x}\, e_0(\bar{m},\bar{x})$ with e_0 free of quantifiers

and suppose $\bar{n}$ are elements of N satisfying $e_0(\bar{m},\bar{n})$. We will complete the diagram:

$$\text{(D)} \qquad \begin{array}{ccc} \langle \bar{m},\bar{n} \rangle & \overset{f}{\dashrightarrow} & M \\ & \nwarrow \quad \nearrow & \\ & \langle \bar{m} \rangle & \end{array} \quad ;$$

then evidently $e_0(\bar{m},\bar{n})$ holds in M, so $e(\bar{m})$ holds, as desired.

We may assume without loss of generality that in Diagram (D) $\bar{n}$ consists of a single element n of N (because the following construction can be iterated). Let M_1 be an injective hull of $\langle\bar{m}\rangle$ and let N_1 be an injective hull of $\langle\bar{m},n\rangle$. Since M and N_1 are injective we may embed M_1 in both M and N_1. Since M_1 is injective we can find direct sum decompositions:

$$M = M_1 \oplus M'$$

$$N_1 = M_1 \oplus N_1'.$$

In particular we may decompose $n = n_1 + n_1'$ with n_1 in M_1 and n_1' in N_1'. We will complete the following diagram:

$$(D') \qquad \begin{array}{ccccc} \langle\bar{m},n\rangle & \longrightarrow & M_1 \oplus \langle n_1'\rangle & \dashrightarrow & M \\ & \nwarrow & & \nearrow & \\ & & \langle\bar{m}\rangle \longrightarrow M_1 & & \end{array}$$

(we have omitted some relevant arrows). In particular this will give us (D).

It clearly suffices to find a copy of $\langle n_1'\rangle$ in M', and we are now in a position to do this rather easily. The relevant facts are the following:

1. M contains a family $\{\langle x\rangle$: x varies over a subset X of $M\}$ of $\kappa(R)$ submodules isomorphic to $\langle n_1'\rangle$, such that the intersection $\langle x_1\rangle \cap \langle x_2\rangle$ of any two such submodules is (0).
2. M_1 can be taken to have cardinality $< \kappa(R)$.

Now let π be the projection from M to M' and let K_x be the kernel of π restricted to $\langle x\rangle$ (for x in X). Then the kernels K_x form a family of $\kappa(R)$ disjoint (i.e. $K_x \cap K_y = (0)$) submodules of M_1. It follows that at least one kernel K_x is (0), and hence π maps at least one $\langle x\rangle$ isomorphically into M'. This produces the desired copy of $\langle n_1'\rangle$ in M', and completes the argument.

<u>Proof of Theorem 8 (Sufficiency)</u>:

If R is coherent then the class of fat $\aleph_0$-injectives is first order axiomatizable; let M_R^* be any axiom system whose models are precisely the fat $\aleph_0$-injectives. We claim that M_R^* is the model companion of M_R. In other words we claim:

An R-module M is existentially complete iff M is fat and $\aleph_0$-injective.

This is a direct consequence of Theorem 14. We remarked there that an existentially complete module is always fat and $\aleph_0$-injective. For the converse, assuming R is coherent consider a fat $\aleph_0$-injective M. M is an elementary substructure of a k(R)-saturated module M' to which Theorem 14 applies (why is M' still fat $\aleph_0$-injective?). Thus M' is existentially complete and therefore M is also existentially complete.

Theorem 12 played a useful role in the proof of Theorem 14. Variants of injective hulls will be helpful in §5 and in Chapter VI §2. We now give the promised proof of Theorem 12.

<u>Proof of Theorem 12</u>:

In the course of this proof the only notion of injectivity which will concern us is the formulation contained in Theorem 5.4. Indeed, we will construct the injective hull of M by adjoining solutions of systems of equations:

(E) $xr = m_r$ (r in S).

As a preliminary remark we notice that M is embeddable in at least one injective module M'. The argument for this is the same one used to build a κ-saturated or an existentially complete structure. One adjoins points systematically until no more are needed (Chapter III §1). Notice that any system of equations (E) involves at most card(R) different parameters from M.

We now describe the construction of an injective hull $\overline{M}$ by transfinite induction. We want to construct a sequence of modules M such that:

1. $M_0 = M$.

2. For δ a limit ordinal:

   $M_\delta = \bigcup_{\alpha<\delta} M_\alpha$

3. If M_α is not injective then $M_{\alpha+1}$ is generated over M_α by a solution to a system of equations (E) which is not solved in M . Furthermore the system of equations (E) is to be <u>maximal</u> in the sense that no system of equations (E') properly containing (E) is solvable over M .

Given M_α, the construction of $M_{\alpha+1}$ involves an element of choice in two ways:

i. A maximal system of equations (E) is to be chosen.

ii. A particular solution of (E) is to be adjoined to M_α.

The freedom referred to in (ii) is purely illusory; in fact one sees easily that if (E) is maximal, then the structure of $M_{\alpha+1}$ over M_α is determined by (E).

We now have to verify that the transfinite induction described eventually constructs an injective module $\overline{M} = M_\alpha$ (for large α) and that $\overline{M}$ is in fact an injective hull of M.

The same argument will be used to verify both claims. Fix an injective N containing M. Notice that the construction outlined in 1-3 may be carried out inside N. (Here we take advantage of the fact that modules can be amalgamated.) In particular the process terminates (at worst, when $M_\alpha = N$), and produces a copy of $\overline{M}$ inside N.

§4. Coherent Rings.

Clearly the material of §§1-2 remains woefully incomplete without an algebraic description of coherent rings (Definition 7). This section is devoted to filling that gap. We recall one of the definitions of $\aleph_0$-injectivity:

M is $\aleph_0$-injective iff for every finite system of equations:

(E) $xr = m_r$ (m_r in M, r varies over a finite set F)

which is solvable in some extension of M, (E) is already solvable in M.

Lemma 15. Let (E) as above be a finite system of equations defined over an R-module M. Then the following are equivalent:

1. (E) is solvable in an extension of M.
2. For any elements $\bar{a}_r \subseteq R$, if $\sum r a_r = 0$ then $\sum m_r a_r = 0$.

Proof: Let $I = \sum rR$. Consider the condition:

3. The map $r \longmapsto m_r$ extends to a homomorphism $h: I \longrightarrow M$.

It is easily checked that 1 => 2 => 3.

3 => 1: Consider the commutative diagram:

$$\begin{array}{ccc} I & \xrightarrow{h} & M \\ {\scriptstyle i}\downarrow & & \downarrow{\scriptstyle i'} \\ R & \xrightarrow[h']{} & M \oplus_I R \end{array} ;$$

it is easily checked that i' is an embedding and $h'(1)$ is a solution of (E).

In spite of its simplicity this lemma contains all that is needed for a description of coherent rings.

Theorem 16. The following are equivalent:

1. R is coherent.
2. For any module homomorphism $h: R^n \longrightarrow R$, the kernel of h is finitely generated.

Proof:

2 => 1: Fix $r_1,\ldots,r_n$ in R. We will axiomatize the class of $\aleph_0$-injective R-modules by axioms of the general form:

($A_{\bar{r}}$) "For all $m_1,\ldots,m_n$ in M, there is an x satisfying $xr_i = m_i$ ($i \leq n$) iff ...."

We must replace the ellipsis by a first order formalization of the predicate:

Solve($\bar{m}$) = " The system of equations $(xr_i = m_i)$ is solvable in some extension of M".

According to Lemma 15, if we introduce the map $h: R^n \longrightarrow R$ defined by:

$h(\delta_{ij}) = r_i$ (δ_{ij} = the jth canonical basis vector)

then Solve($\bar{m}$) is equivalent to:

(*) For all $\bar{a}$ in ker(h) $\Sigma m_i a_i = 0$.

When ker(h) is finitely generated, then (*) may be replaced by a first order formula, as desired.

1 => 2. Suppose M_R^* is the model companion of M_R and that $h:R^n \longrightarrow R$ is a homomorphism. We will show that the kernel of h is finitely generated.

We let $e_j = (\delta_{ij})$ be the jth canonical basis vector in R^n.

Extend the language of R-modules by adjoining n constants $m_1,\dots,m_n$. Then the theory:

(T) $M_R^* \cup \{"\Sigma m_i a_i = 0": \bar{a} \in \ker(h)\}$

proves:

(A) "$\exists x \; xr_i = m_i \; (i \leq n)$"

Hence there are finitely many $\bar{a}^1,\dots,\bar{a}^k$ in ker(h) such that the theory

(T_o) $M_R^* \cup \{"\Sigma m_i a_i^j = 0": j \leq k\}$

proves (A). We claim that $\bar{a}^1,\dots,\bar{a}^k$ generates ker(h). To see this, consider the module:

$M = R^n/\langle\bar{a}^1,\dots,\bar{a}^k\rangle$, and let M' be a model of M_R^* containing M.

M' satisfies T_o if we interpret m_i as e_i. Hence M' satisfies (A) with m_i replaced by e_i. Therefore for any $\bar{a}$ in ker(h) we derive $\Sigma e_i a_i = 0$ in M, and this means that $\bar{a}$ is in $\langle\bar{a}^1,\dots,\bar{a}^k\rangle$, as desired.

§5. Existentially Complete Modules.

When the ring R is not coherent the class $\underset{\sim}{E}$ of existentially complete R-modules is not first order axiomatizable. Nonetheless from the point of view of Chapter III the class $\underset{\sim}{E}$ is rather well-behaved. We prove below that $\underset{\sim}{E}_{M_R} = M_R^f = M_R^\infty$. These results will be approached via some model theory of general R-modules. In the absence of existential completeness we will have fewer injective modules to play with, so as a substitute we develop the notions of purity and pure-injectivity. (In order to use methods like those of the preceding sections we need to know when a submodule M_1 of a module M_2 is a direct summand of M_2.)

Definition 17. If $M_1 \subseteq M_2$ are R-modules we say that M_1 is pure in M_2 (and we write $M_1 \subseteq_p M_2$ or $M_1 \xrightarrow{p} M_2$) iff:

for all $\bar{m} \subseteq M_1$ and for all matrices A over R

if M_2 satisfies $\exists \bar{x}\ \bar{x}A = \bar{m}$

then M_1 satisfies $\exists \bar{x}\ \bar{x}A = \bar{m}$.

We adapt Theorem 5 to the context of R-modules and pure embeddings.

Theorem 18. Let M be an R-module. Then the following are equivalent:

1. M is a direct summand of every pure extension of M.
2. Any diagram of the following form can be completed:

$$(D)\qquad \begin{array}{ccc} M_1 & \xrightarrow{p,1-1} & M_2 \\ h\downarrow & \swarrow & \\ M & & \end{array} .$$

Proof: 2 => 1: as previously.

1 => 2: We consider the following commutative diagram:

$$\begin{array}{ccc} M_1 & \xrightarrow{\ i\ } & M_2 \\ h\downarrow & & \downarrow h' \\ M & \xrightarrow[\ i'\ ]{} & M \oplus_{M_1} M_2 \end{array} .$$

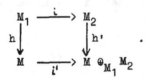

One verifies by diagram chasing that the pure embedding i induces a pure embedding i' (this is worth carrying out). Then we can apply condition 1 to get a projection map $\pi: M \oplus_{M_1} M_2 \longrightarrow M$. The map $\pi h'$ completes diagram (D).

<u>Definition 19</u>. An R-module M is <u>pure-injective</u> iff M satisfies conditions 18.1-2.

It is a pleasant fact that pure-injectives exist in profusion.

<u>Theorem 20</u>. If M is $\operatorname{card}(R)^+$-saturated then M is pure-injective.

<u>Proof</u>: We use a lemma whose proof is left as exercise 11:

<u>Lemma 21</u>. Let $M_1 \subseteq_p M_2$ and let m be an element of M_2. Then there is a pure submodule M_1' of M_2 containing M_1 and m such that M_1' is generated over M_1 by at most $\operatorname{card}(R) + \aleph_0$ elements.

(The point is of course that there may be <u>many</u> equations of the form $\bar{x}A = \bar{m}_1 + \bar{m}_1'$ where $\bar{m}_1$ varies over sequences in M_1 and $\bar{m}_1'$ varies over elements adjoined to M_1 in the process of constructing M_1' - but it is not necessary to treat these equations individually.)

Returning to the proof of Theorem 20, we consider a diagram

(D) 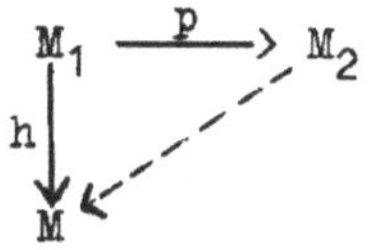

where M is $\operatorname{card}(R)^+$-saturated. By Lemma 21 we might as well assume that M_2 is generated over M_1 by at most card(R) elements (for the general case we can iterate the following argument).

Let therefore A be a generating set of M_2 over M_1. We seek elements x_a (a in A) in M satisfying the following conditions:

(C) $\sum_{a \in A} x_a r_a = h(m_1)$ whenever $\{r_a : a \in A\} \subseteq R$, all but finitely many r_a equal 0, m_1 is in M_1, and $\sum a r_a = m_1$.

Once we find the elements x_a we can determine a map $h': M_2 \longrightarrow M$

extending h by: $h(a) = x_a$.

Since M is $\mathrm{card}(R)^+$-saturated, we need only consider finitely many conditions of the form (C). In this case we may restrict our attention to a finite set $A_o \subseteq A$ and write (C) more concisely as:

(C') $\bar{x}T = h(\bar{m}_1)$ where T is a matrix over R and $\bar{a}T = \bar{m}_1$.

Here $\bar{a}$ varies over A_o and $\bar{x}$ varies over $\{x_a : a \in A_o\}$. Now using the purity of M_1 in M_2 we may solve:

$\bar{y}T = \bar{m}_1$

in M_1, and then set $\bar{x} = h(\bar{y})$, solving (C'). This completes the proof.

From Theorem 20 we see in particular that every R-module admits a pure embedding into some pure-injective module. For our present purposes we will need to refine this observation substantially. The main algebraic ingredient in the theorems of this section will be the following:

<u>Theorem 22</u>. Every R-module M has a <u>pure-injective hull</u> $\bar{M}$ characterized up to isomorphism over M by the following two conditions:

1. $\bar{M}$ is pure-injective, $M \subseteq \bar{M}$

2. Any diagram of the following form can be completed by a pure embedding:

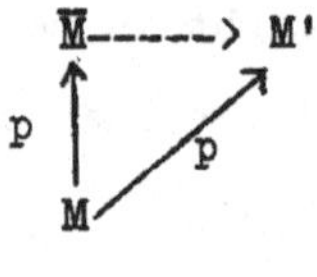

(M' is any pure-injective extension of M).

In addition $\bar{M}$ satisfies the following minimality condition:

3. If $M \subseteq M' \subseteq \bar{M}$ with M' pure-injective then $M' = \bar{M}$.

<u>Notation</u>. If M is an R-module, $\bar{M}$ will denote its pure-injective hull. We remark that the proof that $\bar{M}$ is determined up to isomorphism by 22.1-2 depends on 22.3.

The proof of Theorem 22 will mimic the proof of Theorem 12 (existence of injective hulls). The key idea in that proof was to make systematic use of Theorem 5.4. Hence the first order of business will be to extend Theorem 18 by developing an analog of Theorem 5.4. This is worth doing for its own sake, as the resulting characterization of pure-injectives is illuminating from a purely model-theoretic point of view.

Definition 23.

1. An existential formula $e(x)$ with one free variable x is called positive primitive iff $e(x)$ has the following form:

(e) $\exists \bar{y}\ \ \bar{y}T = x\bar{r} + \bar{m}$

where $\bar{y}$ is some k-tuple of variables, T is some $k \times l$ matrix over R, $\bar{r}$ is an l-tuple of elements of R, and $\bar{m}$ is an l-tuple of constants (denoting parameters from some module M.

In other words, a positive primitive statement says that various systems of equations involving the element x are solvable.

2. A set P of positive primitive formulas $e(x)$ defined over the R-module M is called p-consistent over M iff the following equivalent conditions are satisfied (Exercise 12):

(C1) Every finite subset of P is realized in M.

(C2) P is realized in a pure extension of M.

3. If $m \in M_1 \subseteq M_2$ define the type $P(m,M_2)$ of m in M_2 over M_1 as the set of pos. prim. formulas $e(x)$ defined over M_1 and true of m in M_2 (i.e. M_2 satisfies $e(m)$).

With this terminology we have a natural characterization of pure injective modules along the lines of Theorem 5.4.

Theorem 24. Let M be an R-module. Then the following are equivalent:

1. M is pure-injective.
2. Every type P which is p-consistent over M is realized in M.

Proof: 1 => 2. If P is p-consistent over the pure-injective module M, realize P by an element m of a pure extension M' of

M, and let $\pi: M' \longrightarrow M$ be a projection map (since M is a direct summand of M'). Then one sees easily that πx realizes P in M.

2 => 1. We are asked to complete a diagram:

(D1) 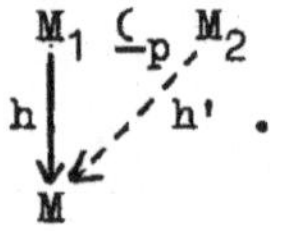

We consider instead a diagram

(D2) 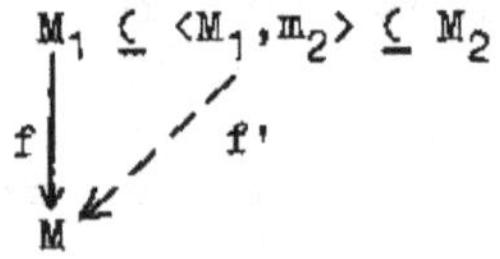

where m_2 is any element of M_2 and f is type-preserving in the following sense: for any positive primitive sentence

$\exists \bar{y}\ \ \bar{y}T = \bar{m}_1 \quad (\bar{m}_1 \subseteq M_1)$

true in M_2, the corresponding sentence

$\exists \bar{y}\ \ \bar{y}T = f(\bar{m}_1)$

is true in M.

Claim. We can complete (D2) via a type-preserving homomorphism f'.

Notice first that if the claim is granted then it is easy to complete (D1). Namely we apply (D2) repeatedly to extend the domain of h to all of M_2, noting that to begin with h is itself type-preserving (a fact which follows at once from the purity of M_1 in M_2).

We now verify our claim. We seek an element x of M satisfying all of the following conditions in M:

(C) $\quad \exists \bar{y}\ \ \bar{y}T = x\bar{r} + f(\bar{m}_1) \quad (\bar{m}_1 \subseteq M_1,\ \ M_2 \text{ satisfies } \exists \bar{y}\ \ \bar{y}T = m_2\bar{r} + \bar{m}_1)$

(We will then be able to set $f'(m_2) = x$.)

Our assumption (2) on M amounts to the following: we may restrict our attention to a finite set of conditions (C). But evidently a finite set of such conditions is equivalent to a single condition (C), which may be written in the form:

(C') $\exists \bar{y}\ \ \bar{y}T - x\bar{r} = f(\bar{m}_1)$, where by assumption

(*) $\exists \bar{y}\ \ \bar{y}T - m_2\bar{r} = \bar{m}_1$ is true in M_2.

The satisfiability of (C') follows immediately from (*) and the fact that f is type-preserving. Namely we have:

(**) $\exists x, \bar{y}\ \ \bar{y}T - x\bar{r} = \bar{m}_1$ is true in M_2.

Hence:

(C") $\exists x, \bar{y}\ \ \bar{y}T - x\bar{r} = f(\bar{m}_1)$ is true in M

and (C') follows. This concludes our syntactical diagram chase.

<u>Remark</u>. Theorem 24 makes Theorem 20 transparent.

We are now prepared for the proof of Theorem 22. We will rely heavily on condition 24.2.

<u>Proof of Theorem 22</u>:

We construct the pure-injective hull $\bar{M}$ of M by realizing various p-consistent types over M. Define by transfinite induction:

1. $M_0 = M$.
2. $M_\delta = \bigcup_{\alpha<\delta} M_\alpha$ for δ a limit ordinal.
3. If M_α is pure-injective let $M_{\alpha+1} = M_\alpha$. Otherwise choose a p-consistent type P which is defined over M_α, is not satisfied in M_α, and is <u>maximal</u> in the sense that any proper extension of P is not p-consistent. Let $M_{\alpha+1}$ be generated over M by a realization of P.

We note that just as in the proof of Theorem 12, the isomorphism type of $M_{\alpha+1}$ over M_α is determined by P. As we have remarked, it follows from Theorem 22 that M is a pure submodule of <u>some</u> pure-injective M'. Therefore the arguments used in Theorem 12 apply to prove:

i. The transfinite induction outlined above eventually produces a pure-injective module $\bar{M} = M_\alpha$ (for large α).

ii. $\bar{M}$ is a pure-injective hull of M (i.e. $\bar{M}$ satisfies 22.1-2).

We must now verify that $\overline{M}$ contains no proper pure-injective submodule M' containing M. Suppose that:

$M \subseteq M' \subseteq \overline{M}$ and M' is pure-injective.

In particular $\overline{M} = M' \oplus M''$ for some complementary R-module M''; let $\pi: \overline{M} \longrightarrow M'$ be the corresponding projection map. If $M'' = (0)$ then $\overline{M} = M'$ as desired. Otherwise let α be minimal such that

$M_{\alpha+1} \cap M'' \neq (0)$.

Choose a nonzero element m'' of $M'' \cap M_{\alpha+1}$. Notice that π maps M_α isomorphically into M', and $\pi m'' = 0$. We know that $M_{\alpha+1}$ is generated over M_α by a solution m to a maximal p-consistent type P, and we may write

$m'' = mr + m_\alpha$ for suitable r in R, m_α in M_α.

Consider the type

$P' = P \cup \{xr + m_\alpha = 0\}$.

Then P' properly extends P, and is easily seen to be p-consistent with M, contradicting the maximality of P (just notice that P' is satisfied by πm over $\pi[M_\alpha]$). This contradiction shows that $M'' = (0)$ and completes the argument.

We leave the proof that $\overline{M}$ is unique up to isomorphism to the reader.

We need one more general observation concerning pure-injective hulls.

Theorem 25. Let M_1, M_2 be R-modules. Then $\overline{(M_1 \oplus M_2)} = \overline{M}_1 \oplus \overline{M}_2$. (This means that $\overline{M}_1 \oplus \overline{M}_2$ is an injective hull of $M_1 \oplus M_2$ with respect to the natural embedding $M_1 \oplus M_2 \longrightarrow \overline{M}_1 \oplus \overline{M}_2$.)

Proof: Pure diagram chasing, using conditions 22.1-2. (Exercise 13)

Theorem 22 becomes a useful tool for studying the model theory of modules when used in conjunction with Theorem 20. Whenever we saturate a module or a collection of modules sufficiently we get pure-injectivity "for free". We use Theorem 22 to prove the following purely algebraic theorem; then using Theorem 20 we will derive an immediate model-theoretic conclusion.

Theorem 26. Suppose M_1, M_2 are pure-injective and we have pure-embeddings:

(*) $M_1 \xrightarrow{p} M_2 \xrightarrow{p} M_1$.

Then $M_1 \simeq M_2$.

Proof: Use the pure-injectivity of M_1, M_2 in conjunction with (*) to write:

(A) $M_1 = N_1 \oplus M_2$

(B) $M_2 = N_2 \oplus M_1$.

Applying (B) to (A), and iterating this process, one obtains:

$M_1 \underset{p}{\supseteq} (N_1 \oplus N_2 \oplus N_1 \oplus \ldots) = N_1^{(\aleph_0)} \oplus N_2^{(\aleph_0)}$.

Hence $M_1 \underset{p}{\supseteq} \overline{(N_1^{(\aleph_0)} \oplus N_2^{(\aleph_0)})} = \overline{(N_1^{(\aleph_0)})} \oplus \overline{(N_2^{(\aleph_0)})}$.

We may write:

(A') $M_1 = \overline{(N_1^{(\aleph_0)})} \oplus \overline{(N_2^{(\aleph_0)})} \oplus M_1'$,

(B') $M_2 = N_2 \oplus \overline{(N_1^{(\aleph_0)})} \oplus \overline{(N_2^{(\aleph_0)})} \oplus M_1'$.

(B') comes from (A') using (B). From (A'),(B') and Theorem 25 we see that $M_1 \simeq M_2$.

The following is essentially a corollary:

Theorem 27. $Th(\underset{\sim}{E}_{M_R}) = Th(M_R^f) = Th(M_R^\infty)$.

(For the notation consult Chapter III.)

Proof: It suffices to show that the theory of $\underset{\sim}{E}_{M_R}$ is complete, since it is contained in the other two theories. In other words, assume M_1, M_2 are existentially complete; we must show that M_1, M_2 are elementarily equivalent. After some minor fol-de-rol we will see that this is essentially a special case of Theorem 26.

Let M_3 be an existentially complete module containing M_1, M_2 (for example take any existentially complete extension of $M_1 \oplus M_2$). We will show that $M_1 \equiv M_3$ (hence by symmetry $M_2 \equiv M_3$ and thus $M_1 \equiv M_2$).

By Theorem III. 10we can complete the diagram:

(D) $M_1 \xrightarrow{f} M_3 \overset{g}{-\!\!-\!\!>} M_1'$

using an elementary extension M_1' of M_1. If we saturate the following structure:

$\langle M_1, M_3, M_1'; f, g\rangle$

we obtain a diagram

(D*) $M_1^* \xrightarrow{f^*} M_3^* \xrightarrow{g^*} M_1^*$

where M_1^* is a saturated elementary extension of M_1 and we have replaced $M_1'^*$ by M_1^* since they are isomorphic (Theorem 0.14).

Now in (D) the maps are pure because M_1, M_3 are existentially complete. Since (D*) is an elementary extension of (D), the embeddings f^*, g^* are pure. We may take M_1^*, M_2^* to be at least $\mathrm{card}(R)^+$-saturated, hence pure-injective. We thus recognize (D*) as a special case of 26(*). Thus $M_1^* \simeq M_3^*$. In particular:

$M_1^* \equiv M_3^*$.

It follows that $M_1 \equiv M_3$, as desired.

We can immediately strengthen Theorem 27, using the same methods at somewhat greater length:

Theorem 28.

1. $\underset{\sim}{E}_{M_R} = M_R^f$.
2. More generally, if $M_1 \subseteq_p M_2$ and $M_1 \equiv_2 M_2$ (Chapter 0 §6) then $M_1 \prec M_2$.

Remarks 29.

1. We rephrase 28.1 in the form we will actually prove:

(1) If M_1 is existentially complete, $M_1 \subseteq M_2$, and $M_1 \equiv M_2$, then $M_1 \prec M_2$.

By Theorem III. 54 this statement is equivalent to 28.1. Furthermore it is clear that this statement is generalized by 28.2 (recalling that $\equiv_2$ is a weak version of $\equiv$).

2. We prove 29(1) (and hence 28.1) below. We leave 28.2 for the reader (Exercise 14).

Proof of 29(1):

Begin with the diagram:

(D) $M_1 \xrightarrow{i} M_2$. The inclusion map i is pure since M_1 is existentially complete. Saturating the diagram (D) sufficiently we obtain:

(D*) $M_1^* \xrightarrow{i^*} M_2^*$; i^* is pure, M_1^*, M_2^* are pure-injective, $M_1^* \simeq M_2^*$.
(The last statement is a consequence of the assumption: $M_1 \equiv M_2$).

We will show that $M_1^* \prec M_2^*$, and hence $M_1 \prec M_2$.

Since this proof will tend to run on a bit, we now <u>drop all asterisks</u> from our notation (set $M_i^* = M_i$, $i^* = i$). Then we have:

i. $M_1 \simeq M_2$

ii. $M_2 = M_1 \oplus M$ for some complementary submodule M.

(This is a reformulation of (D*).) Notice that M is pure-injective (Exercise 16).

We want to show that the inclusion $M_1 \subseteq M_2$ is an elementary embedding. In other words, fix an arbitrary n-tuple $\bar{m} \subseteq M_1$; we must show that $\langle M_1, \bar{m}\rangle \equiv \langle M_2, \bar{m}\rangle$. We will in fact show something stronger:

Claim. The following diagram can be completed by an isomorphism f:

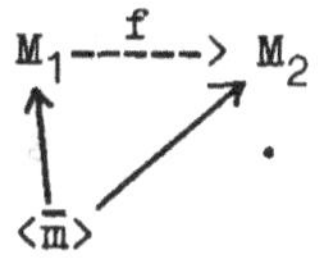

Verification. From i,ii it follows easily by induction that

$M_1 \simeq M_1 \oplus M^n$ for all n.

Suppose for the moment that

(*) $M_1 \simeq M_1 \oplus M^{(\aleph_0)}$ (an unlikely circumstance).

In this case the embedding $M_1 \longrightarrow M_2$ is just the canonical inclusion $M_1 \oplus M^{(\aleph_0)} \longrightarrow M_1 \oplus M^{(\aleph_0+1)}$. The claim above is then evidently correct.

We will replace (*) by something equally useful and more plausible. For this purpose we will take M_1 saturated in cardinality at least $\text{card}(R)^{++}$ (a harmless assumption). We study the modules

$N_\alpha = M_1 \oplus (M^{(\alpha)})^{-}$.

In particular $N_0 = M_1$, $N_1 = M_2$.

Our main claim is: for all $\alpha < \text{card}(M_1)^+$ we have

(**) $M_1 \simeq N_\alpha$.

We will also see that for $\alpha = \text{card}(R)^+$, (**) is as useful as (*).

First we mention two relatively technical points, whose proofs are left to the reader as Exercise 15:

(A) $\{N_\alpha\}$ is a pure-directed system. By this we mean that there are pure embeddings $i^\alpha_\beta : N_\alpha \xrightarrow{p} N_\beta$ for $\alpha < \beta$ satisfying $i^\beta_\gamma \circ i^\alpha_\beta = i^\alpha_\gamma$.

(B) $N_\delta = (\lim\limits_{\alpha<\delta} N_\alpha)^{-}$ for all limit ordinals δ.

At first glance (A) requires no proof, since the canonical embeddings $j^\alpha_\beta : M^{(\alpha)} \xrightarrow{p} M^{(\beta)}$ induce embeddings $i^\alpha_\beta : N_\alpha \xrightarrow{p} N_\beta$. However the i^α_β are not canonical, so some further care is required. One use of (A) is to make sense out of (B).

Now we return to the proof of (**), the central point in this argument. If $\alpha = o$ then (**) is trivial. Given $M_1 \simeq N_\alpha$, we compute $N_{\alpha+1} \simeq N_\alpha \oplus M \simeq M_1 \oplus M \simeq M_2 \simeq M_1$, as desired. We must now verify (**) in the case of a limit ordinal $\delta < \text{card}(M_1)^+$, assuming $M_1 \simeq N_\alpha$ for $\alpha < \delta$.

According to Theorem 26 we need only find pure embeddings $N_\delta \xrightarrow{p} M_1$, $M_1 \xrightarrow{p} N_\delta$. Now $i^o_\delta : M_1 \xrightarrow{p} N_\delta$, so only the embedding $N_\delta \xrightarrow{p} M_1$ is lacking. Since $N_\delta = (\lim\limits_{\alpha<\delta} N_\alpha)^{-}$, it suffices to carry out the following task:

(T) Find a map: $\lim\limits_{\alpha<\delta} N_\alpha \xrightarrow{p} M_1$.

Why is this possible? The short answer is that this is a special case of results in Chapter 0 §6 (we will elaborate). For the reader's convenience we also include a direct (though somewhat idiosyncratic) proof of the existence of the desired map. It will be convenient to set $L = \lim\limits_{\alpha<\delta} N_\alpha$.

(T)- First Version:

Since $M_1 \simeq N_\alpha$ for $\alpha < \delta$ it follows that $M_1 \prec_2 L$. Hence there is an embedding $L \prec_1 M_1'$ with $M_1' \equiv M_1$. Then $L \subseteq_p M_1'$ and $M_1' \xrightarrow{\equiv} M_1$ (Corollary 0.14).

The composition $L \xrightarrow{p} M_1' \xrightarrow{p} M_1$ produces the desired pure embedding.

(T)- Second Version:

We collect everything we know about L in one big structure. Namely we define:

$A = \langle L, M_1, \delta; \ <, E, f, \{c_l : l \in L\}, \{c_\alpha : \alpha \in \delta\}\rangle$. Here we take:

1. $<$ is the usual ordering of δ.
2. E is the following relation: Exy iff $x \in L$, $y \in \delta$, $x \in N_y$.
3. f(x,y) is a function of two variables such that for $y \in \delta$, $f(\cdot, y): N_y \simeq M_1$.
4. The c's are constants naming various elements.

Now we replace A by a saturated elementary extension A* of A such that card(A*) = card(M_1). Then A* has the following form:

$\langle L^*, M_1, \delta^*; \ <, E, f, \{c_l\}, \{c_\alpha\}\rangle$.

We have identified M_1^* with M_1 (Theorem 0.15), and left the asterisks off the relation symbols.

Fix an element y in A* realizing the type:

(p) $y > \alpha \quad (\alpha < \delta)$; (we may assume this type is realized).

Define $N_y^* = \{x \in L^* : Exy\}$. Define $f_y = f(\cdot, y)$.

Then one verifies easily:

I. $f_y : N_y^* \simeq M_1$.

II. $L \subseteq_p N_y^*$.

Combining i,ii we have the desired embedding: $L \xrightarrow{p} M_1$.

Thus one way or another we may carry out (T) and complete the proof of (**).

Now taking $\delta = \text{card}(R)^+$ we have:

$M_1 \simeq N_\delta = (\varinjlim_{\alpha<\delta} N_\alpha)^{-} = \varinjlim_{\alpha<\delta} N_\alpha$.

The last equation follows from Theorem 24 or Theorem 20 (each N_α is saturated). Thus the inclusion $M_1 \subseteq M_2$ may be viewed as

$$\text{(I)} \quad \lim_{\alpha<\delta} M \oplus (M^{(\alpha)})^{-} \subseteq \lim_{\alpha<\delta} M \oplus (M^{(\alpha+1)})^{-}$$

using ii above.

It is evident that the two sides of this last inclusion are isomorphic over any finitely generated submodule, so our claim is verified.

Corollary 30. $\underset{\sim}{E}_{M_R} = M_R^f = M_R^\infty$.

Proof: It follows from Corollary III.49 and the model-completeness of M_R^f that the first equality implies the second.

§6. Notes.

§§1-4 are based on |21|. The material in §5 is due to Sabbagh |41| and Fisher |24|. Pure injective hulls are studied in |53|.

Exercises.

§1.

1. Prove that every abelian group is contained in a divisible group, and that a direct summand of a divisible group is divisible.
2. Let A be divisible. Show that any diagram:

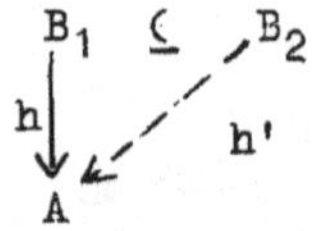

can be completed by a homomorphism h' (we do not assume h is 1-1). Taking $B_1 = A$, conclude that A is a direct summand of B_2. (This exercise, generalized to arbitrary modules, is part of Theorem 5: "5.4 => 5.2".)

§2.

3. Consider a diagram

$$\begin{array}{ccc} M & \xrightarrow{f} & M_1 \\ \downarrow & & \downarrow \\ M_2 & \xrightarrow[f']{} & M_1 \oplus_M M_2. \end{array}$$

If f is 1-1 prove that f' is 1-1.

4. Prove that a direct sum of $\aleph_0$-injective modules is $\aleph_0$-injective.

5. Suppose the class of injective R-modules is closed under formation of infinite direct sums. Prove that R is noetherian.

6. Suppose that M is injective and N is a direct summand of M. Show that N is injective.

§3.

7. Verify Lemma 10.

8. Prove that principal ideal domains are coherent and give a simple axiom system for M_R^*.

9. Generalize ($\oplus$) (following Proposition 2) to injective modules.

§4.

10. Let R be a ring. Prove that the following are equivalent:

 i. The class of injective R-modules is first order axiomatizable.

 ii. R is noetherian.

(Suggestion: assuming (i) prove that R is coherent and that $\aleph_0$-injectives are injective; then apply earlier exercises.)

§5.

11. Verify Lemma 21.

12. Verify the equivalence of conditions 23(C1-2).

13. Verify Theorem 25.

14. Verify that the proof of 29(1) also proves 28.2.

15. Show by induction on β that:

(A) $\{N_{\alpha+1} : \alpha < \beta\}$ is a pure-directed system.

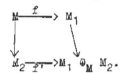

(B) $N_\beta = (\lim_{\alpha<\beta} N_{\alpha+1})^{-}$.

16. Show that a direct summand of a pure-injective module is pure-injective.

VI. Complete Theories of Abelian Groups

Introduction.

The main result in this chapter will be a classification of all complete theories of abelian groups, that is to say of all theories of the form Th(A) where A is an abelian group. The isomorphism types of finite abelian groups are classified by familiar numerical invariants, but this is not the case for isomorphism types of infinite abelian groups (cf. |19|); it is an agreeable circumstance that complete theories of abelian groups can be classified by simple numerical invariants, as we will prove. Since a finite structure is determined up to isomorphism type by its theory, the classification of complete theories of abelian groups includes the structure theorem for finite abelian groups.

Our goal therefore is the following:

(*) Classify all abelian groups up to elementary equivalence.

This is equivalent to the following (modulo set-theoretic equivocation as in Chapter 0):

(**) Classify all saturated abelian groups up to isomorphism.

As a lengthy exercise the energetic reader might peruse |27| and then accomplish (**). We give a self-contained account of (**) in §1. §2 contains among other things a proof that groups which appear to be saturated actually are (regardless of one's set-theoretic ideology). Using the material of §§1-2 a number of questions in the model theory of abelian groups become trivial- or at least tractable. Some of these are found in the exercises.

§1. Structure of Saturated Abelian Groups.

First order statements in the language of abelian groups are relatively uncomplicated. Given an element a of an abelian group A, the only interesting statements one can express concerning a are:

1. torsion statements ("na = 0" for various n).

2. divisibility statements ("n divides a" for various n).

Of course if n divides a then a/n is usually not uniquely determined. (From the point of view of Chapter V torsion statements describe the cyclic $\underset{\sim}{Z}$-module ⟨a⟩ and divisibility statements control the pure-injective hull(s) of ⟨a⟩)

The foregoing is worth bearing in mind as a heuristic principle; a rigorous version of it will be implicit in §2.

We are going to exhibit an arbitrary saturated abelian group as a plump version of a direct sum of certain basic abelian groups (the precise statement is Theorem 2 below). The basic abelian groups are:

α. Z/p^n = the integers modulo (p^n) (this is divisible by all integers m relatively prime to p and has p-torsion). To see that Z/p^n has the divisibility property claimed, recall the usual proof that $\langle Z/p,+,\cdot\rangle$ is a field.

β. $Z_p = \underset{\sim}{Z}$ localized at $p = \{a/b \in \underset{\sim}{Q} : (b,p) = 1\}$ (this is divisible by all integers m relatively prime to p and has no torsion). Z_p can also be described as the intersection of the ring of p-adic integers with $\underset{\sim}{Q}$. Equipped with the obvious multiplication Z_p is a ring.

Notice that Z/p^n, Z_p are cyclic Z_p-modules.

γ. $Z/p^\infty = \{z \in C : \text{for some } n \ z^{p^n} = 1\} = Z_p/\underset{\sim}{Z}$ (divisible, with p-torsion).

δ. $\underset{\sim}{Q}$ (divisible with no torsion).

We may also write $Z/p^{\infty} = \varinjlim Z/p^n$ where we use the injections $i_k^n\colon Z/p^n \to Z/p^{n+k}$ defined by: $i_k^n(z) = p^k z$. Alternatively Z/p^{∞} is the divisible hull of Z/p, $\underset{\sim}{Q}$ is the divisible hull of $\underset{\sim}{Z}$.

Non-divisibility.

Torsion.

	p	none
p	$\alpha: Z/p^n$	$\gamma: Z/p^{\infty}$
none	$\beta: Z_p$	$\delta: \underset{\sim}{Q}$

<u>The Basic Abelian Groups</u>

We will perform the following operations on these abelian groups:

1. Direct sum
2. p-adic closure (see below)
3. Direct product.

In connection with (2) we introduce the following terminology and notations.

<u>Definition 1</u>. Let A be an abelian group, a an element of A. Define:

$\mathrm{ord}_p(a)$ = the greatest n such that $p^n | a$ in A (or ∞).

$|a|_p = p^{-\mathrm{ord}_p(a)}$ (or 0 if $\mathrm{ord}_p(a) = \infty$).

The topology induced by the seminorm $|\ |_p$ on A is called the p-topology. A is p-Hausdorff iff $|a|_p = 0 \Rightarrow a = 0$ for a in A (in other words, $|\ |_p$ is a norm). If A is p-Hausdorff then A has a p-completion $\bar{A}$ (more explicitly: $\bar{A}^p$) which is again an abelian group (in fact a complete Hausdorff topological abelian group). The topological terminology is very convenient for analyzing divisibility properties of subgroups of A; of course the p-topology has a neighborhood basis of first order definable sets at 0.

The topology determined by all the seminorms $|\ |_p$ (or equivalently by $|\ | = \sum 2^{-p}|\ |_p$) is called the Z-topology. If A is Hausdorff in the Z-topology (i.e. if A contains no infinitely divisible element other than 0) then A is said to be <u>reduced</u>.

The main theorem is the following:

<u>Theorem 2</u>. Let κ be an uncountable cardinal, A a κ-saturated abelian group. Then A is of the form:

$$A = \prod_p \bar{A}_p \oplus D$$

where D is divisible and $\bar{A}_p$ is the completion of a p-Hausdorff Z_p-module A_p. Furthermore A_p,D may be decomposed as follows:

$$A_p = \oplus\sum_n (Z/p^n)^{(\alpha_{p,n})} \oplus Z_p^{(\beta_p)}.$$

$$D = \oplus\sum_p (Z/p^\infty)^{(\gamma_p)} \oplus Q^{(\delta)}.$$

Here $\alpha_{p,n}$, β_p, γ_p are arbitrary finite cardinals or $\geq \kappa$, δ is 0 or $\geq \kappa$, and the notation $B^{(\lambda)}$ denotes the direct sum of λ copies of B (the direct product would be denoted B^λ).

Taken with a grain of salt the converse is also true (there are relations among the $\alpha,\beta,\gamma,\delta$).

<u>Convention</u>. Until Definition 18 $\kappa \geq \aleph_1$, A is a κ-saturated abelian group, $|\ |_p$ is the p-adic seminorm on A, and $|\ | = \sum 2^{-p}|\ |_p$. We will analyze the structure of A in several steps.

<u>Definition 3</u>. $D = \{a \in A: |a| = 0\}$.

$R = A/D$.

<u>Proposition 4</u>. D is saturated and divisible and R is Hausdorff and complete with respect to $|\ |$ (i.e. in the Z-topology).

<u>Proof</u>: We use repeatedly the saturation of A.

<u>D is divisible</u>: for a in D and any integer n, consider the conditions:

(*) $nx = a$; $|x| = 0$.

The second condition is equivalent to infinitely many first order conditions. Because A is saturated and a is in D it is evident that (*) is satisfiable. Thus D is divisible.

That D is saturated follows in similar fashion.

Since D is divisible it is a direct summand of A (cf. Chapter V §§1-2). With a slight abuse of notation we may write $A = D \oplus R$ (the decomposition is not canonical). It is then evident that R is Hausdorff. The completeness of R is an immediate consequence of the saturation of A and the definability of the relevant notions in the p-topology.

Definition 5. $R_p = \{a \in R: |a| = |a|_p\}$ (this is the set of elements of R which are infinitely divisible by every prime other than p).

Proposition 6. $R = \prod_p R_p$. Each R_p is a p-complete p-Hausdorff Z_p-module.

Proof: R_p is closed in R in the Z-topology, hence Z-complete. On R_p the Z-topology and the p-topology coincide, so R_p is p-complete.

Clearly $\oplus\sum R_p \subseteq R$ (the sum is direct since R is Hausdorff). The completion of $\oplus\sum R_p$ in the Z-topology is easily seen to be $\prod_p R_p$, so $\prod_p R_p \subseteq R$.

Fix $a \in R$. We claim there is an $a_p \in R_p$ so that $|a-a_p|_p = 0$. Assuming this fact for the moment let $a' = (a_p) \in \prod_p R_p$. Then $|a-a'|_p = 0$ for all p, hence $a = a' \in \prod_p R_p$.

Thus to complete the argument we need only find an $x = a_p$ satisfying:

(*) "$m|x$" (for all m relatively prime to p)

"$p^n|(a-x)$" (for all n).

It suffices to show that any finite set of conditions of the form (*) is satisfied in A. Without loss of generality we may deal with one condition of the form "$m|x$" and one of the form "$p^n|a-x$".

Simply take $x = sa$ where s is an integer chosen (via the Chinese Remainder Theorem) to satisfy:

$$s \equiv 0 \pmod{m}, \quad s \equiv 1 \pmod{p^n}.$$

<u>Proposition 7</u>. Let D be a divisible group. Then D has the following form for suitable γ_p, δ:

$$D = \oplus\Sigma(Z/p^{\infty})^{(\gamma_p)} \oplus \underset{\sim}{Q}^{(\delta)}.$$

<u>Proof</u>: Let $D_p = \{x \in D\colon \text{for some } n \;\; p^n x = 0\}$. Then $D \supseteq \oplus\Sigma\, D_p$, and the latter group is clearly divisible, hence is a direct summand of D. If we write $D = \oplus\Sigma\, D_p \oplus D_o$ then D_o is torsion free (i.e. has no torsion), and may therefore be construed as a vector space over $\underset{\sim}{Q}$. Thus we may write

$$D_o = \underset{\sim}{Q}^{(\delta)} \quad \text{for some } \delta .$$

We claim now that each D_p has the form:

$$D_p = (Z/p^{\infty})^{(\gamma_p)} \quad \text{for some } \gamma_p.$$

By Zorn's lemma there is a maximal divisible subgroup D_p' of D having the desired form (this could be said more carefully: Exercise 1). D_p' is divisible, hence a direct summand of D_p. Write $D_p = D_p' \oplus D_p''$. Since D_p' is maximal, D_p'' contains no copy of Z/p^{∞}. We claim that this forces $D_p'' = 0$, so that $D_p = D_p'$ has the desired form. We leave this as an exercise (Exercise 2). The relevant facts are:

1. D_p'' is divisible.
2. Every element of D_p'' is annihilated by a power of p. $\dashv$

We now revert to the study of subgroups R_p as described in Proposition 6. In Proposition 7 every group in sight is divisible, which makes it rather easy to obtain direct sum decompositions. The study of R_p requires a little more care. In order to locate direct summands of abelian groups we will use the notion of pure-injectivity as developed in Chapter 5 §5.

<u>Definition 8</u>. Let $A_1 \subseteq A_2$ be abelian groups.

Then A_1 is said to be <u>pure</u> in A_2 iff for every element a of A_1 and every integer n:

n divides a in A_1 iff n divides a in A_2.

The conflict with the terminology of Chapter V §5 is resolved in the next lemma.

Lemma 9. Let $A_1 \_ A_2$ be abelian groups. Then A_1 is pure in A_2 as an abelian group iff A_1 is pure in A_2 as a $\underset{\sim}{Z}$-module.

Proof: Let A_1 be pure in A_2 as abelian groups. We will show that A_1 is pure in A_2 as Z-modules (the converse is trivial).

We may without loss of generality assume that A_2 is finitely generated over A_1, we are concerned only with finite systems of linear equations. Then by Lemma II. 29 A_1 is a direct summand of A_2. Thus A_1 is certainly pure in A_2 as a $\underset{\sim}{Z}$-module.

Proposition 10. Let R_p be a p-complete p-Hausdorff Z_p-module. Then R_p is the p-completion of a diredt sum of cyclic Z_p-modules. Explicitly:

$$R_p = \overline{A}_p; \quad A_p = \oplus\Sigma\, (Z/p^n)^{(\alpha_{p,n})} \oplus Z_p^{(\beta_p)}.$$

Proof: Choose a submodule $A_p \subseteq R_p$ maximal subject to:

1. A_p is pure in R_p.
2. A_p is a direct sum of cyclic Z_p-modules.

(Cf. Exercise 1.)

It is easily seen that $\overline{A}_p$ is pure in R_p. By Lemma 9.1 $\overline{A}_p$ is a direct summand of R_p. Write $R_p = \overline{A}_p \oplus R_p'$. Since A_p is maximal subject to 1,2 it follows that $R_p' = (0)$, for otherwise we may fix an element a in R_p' such that p does not divide a, and then $\langle a\rangle_{Z_p}$ (the corresponding cyclic Z_p-module) is contained in R_p', a contradiction.

Thus $R_p' = 0$, $R_p = \bar{A}_p$, as desired.

We pause to summarize the foregoing.

Proposition 11. Let A be an $\aleph_1$-saturated abelian group. Then A has the form:

(*) $\prod_p [\oplus\Sigma(Z/p^n)^{(\alpha_{p,n})} \oplus Z_p^{(\beta_p)}]^{-p} \oplus \oplus\Sigma(Z/p^\infty)^{(\gamma_p)} \oplus Q^{(\delta)}$.

(In fact (*) characterizes the pure-injective abelian groups.)

Proof: Propositions 4,6,7,10.

Comparing Proposition 11 with the statement of Theorem 2, we conclude that the primary task remaining is to show the first order definability of the relations:

(R) "$\alpha_{p,n} \geq k$", "$\beta_p \geq k$", "$\gamma_p \geq k$".

We also need:

Proposition 12. If A is not of bounded order then $\delta \geq \kappa$.

Proof: Exercise 4.

We will deal in a succession of lemmas with the definability of the relations (R). We will of course consider various properties of A involving torsion and divisiblity. The main complication is the following: for large values of n, Z/p^n resembles both Z/p^∞ and Z_p (compare the chart preceding Definition 1). We should also bear in mind that the factors occurring in Proposition 11 are not canonically constructed, and there is no reason to expect the $\alpha, \beta, \gamma, \delta$ to be invariants of A.

Notation. For B an abelian group and n an integer set

$B\langle n\rangle$ = the set of elements of B annihilated by n.

Convention. We fix a decomposition (*) of the κ-saturated group A.

Lemma 13. $\alpha_{p,n} = \dim_p (p^{n-1}A)\langle p\rangle/(p^nA)\langle p\rangle$. (This factor group has exponent p, evidently, hence has a dimension as a vector space over F_p.)

Proof: In the first place it is clear that:

1. $(p^{n-1}A)\langle p\rangle / (p^nA)\langle p\rangle =$

$$= (p^{n-1}\ \overline{\oplus\Sigma\ (Z/p^k)^{(\alpha_{p,k})}})\langle p\rangle / (p^n\ \overline{\oplus\Sigma\ (Z/p^k)^{(\alpha_{p,k})}})\langle p\rangle.$$

(Use 11(*)). We make use of the formula:

2. $\overline{A_1\oplus A_2} = \overline{A}_1\oplus\overline{A}_2$ (which is already needed for 1),

taking:

3. $A_1 = \underset{k\leq n}{\oplus\Sigma}\ (Z/p^k)^{(\alpha_{p,k})}$

4. $A_2 = \underset{k>n}{\oplus\Sigma}\ (Z/p^k)^{(\alpha_{p,k})}$.

Notice now:

5. $\overline{A}_1 = A_1$, and

6. $(p^{n-1}\overline{A}_2)\langle p\rangle = (p^n\overline{A}_2)\langle p\rangle$ (this last point is rather important, and is worth checking).

From 1-6 it follows rather easily that:

$$(p^{n-1}A)\langle p\rangle / (p^nA)\langle p\rangle = (p^{n-1}A_1)\langle p\rangle / (p^nA_1)\langle p\rangle.$$

But $p^nA_1 = 0$ and $p^{n-1}A_1 = p^{n-1}(Z/p^n)^{(\alpha_{p,n})}$.

The Lemma follows immediately.

The following three lemmas are all concerned with the value of β_p. The sole purpose of Lemma 14 is to establish that β_p is an invariant of A (this is used tacitly in Lemma 16).

Lemma 14. Let T be the torsion subgroup of A (consisting of all elements of A annihilated by some integer). Then

$$\beta_p = \dim_p (A/\ T+pA).$$

Proof: pA contains all copies of $\underset{\sim}{Q}$ or Z/p^∞ in A, as well as

$$\prod_{q\neq p} [\oplus\Sigma(Z/q^n)^{(\alpha_{q,n})} \oplus Z_q^{(\beta_q)}]^{\frown q}.$$

Hence without loss of generality take

$$A = [\oplus\Sigma\ (Z/p^n)^{(\alpha_{p,n})} \oplus Z_p^{(\beta_p)}]^{\frown p}.$$

Let $A_1 = (\oplus\Sigma\ (Z/p^n)^{(\alpha_{p,n})})$. Although $A_1 \subseteq T$ we do not claim $\bar{A}_1 \subseteq T$; we merely claim that $\bar{A}_1 \subseteq T+pA$, and we leave the verification of this fact to the reader.

It follows easily that $\dim_p (A/\ T+pA) = \dim_p(Z_p^{(\beta_p)} / pZ_p^{(\beta_p)})$, and the latter is evidently β_p, as desired.

<u>Lemma 15</u>. $\dim_p (p^{n-1}A / p^n A) = \sum_{k\geq n}\alpha_{p,k} + \beta_p$.

<u>Proof</u>:

Let $A_1 = \oplus\sum_{k<n} (Z/p^k)^{(\alpha_{p,k})}$, $A_2 = \oplus\sum_{k\geq n} (Z/p^k)^{(\alpha_{p,k})}$, and $A_3 = Z_p^{(\beta_p)}$.

Evidently $(p^{n-1}A / p^n A) = (p^{n-1}A_1 / p^n A_1) \oplus (p^{n-1}\bar{A}_2/p^n\bar{A}_2) \oplus \oplus (p^{n-1}\bar{A}_3/p^n\bar{A}_3)$.

On the right hand side the first term vanishes. We claim that in the remaining two terms we may replace $\bar{A}_2, \bar{A}_3$ simply by A_2, A_3. From this the Lemma will follow at once.

Let B be any p-Hausdorff group (e.g. $B = A_2$ or A_3). Any element b of $\bar{B}$ may be written $b = b' + pb''$ with b' in B, b'' in $\bar{B}$. It follows easily that $p^{n-1}\bar{B}/p^n\bar{B} = p^{n-1}B/p^nB$. This completes the argumeht.

<u>Lemma 16</u>. Let p be a fixed prime.

1. If for sufficiently large n $\alpha_{p,n} \equiv 0$ then for large n: $\beta_p = \dim_p (p^n A/p^{n+1}A)$.
2. If there are arbitrarily large n such that $\alpha_{p,n} \neq 0$ then $\beta_p \geq \kappa$.

<u>Proof</u>: For 1, use Lemma 15.

2. Assume there are arbitrarily large n such that $\alpha_{p,n} \neq 0$. It suffices to verify that the condition:

(*) "$Z_p^{(\kappa)}$ is a pure subgroup of A"

is consistent with $Th(A)$. (The condition (*) can be expressed in terms of the realization of a type involving κ distinct variables $\{x_a: a \in Z_p^{(\kappa)}\}$. If this type is consistent with $Th(A)$ then it is realized in A (Chapter 0 §2), which means that β_p is at least κ.)

To see that (*) is satisfiable in A it suffices to check the satisfiability of:

(**) "Z_p^k is a pure subgroup of A"

for every integer k. We leave this to the reader (Exercise 5).

Lemma 17. γ_p = the eventual value of $\dim_p (p^n A\langle p\rangle)$, if this value is finite. Otherwise $\gamma_p \geq \kappa$.

Proof: Like the proof of the previous lemmas (Exercise 6). Note that $\dim_p(p^n A\langle p\rangle)$ is a decreasing function of n, and is therefore eventually constant.

Proof of Theorem 2:

A is a κ-saturated group with $\kappa \geq \aleph_1$. Proposition 11 furnishes a decomposition of A of the desired form; in addition we have claimed that $\alpha_{p,n}, \beta_p, \gamma_p$ are finite or $\geq \kappa$, and that δ is 0 or $\geq \kappa$. These claims are verified in Lemma 13 (in conjunction with the usual saturation argument), Lemma 16, Lemma 17, and Proposition 12 respectively.

At this point we naturally assign the following invariants to an arbitrary abelian group A (not necessarily saturated!):

Definition 18 Let A be an abelian group.

$$\alpha_{p,n}(A) = \begin{cases} \dim_p (p^{n-1}A)\langle p\rangle / (p^n A)\langle p\rangle & \text{if this is finite} \\ \infty & \text{otherwise.} \end{cases}$$

$$\beta_p(A) = \begin{cases} \text{eventual value of } \dim_p p^{n-1}A/p^n A & \text{if finite} \\ \infty & \text{otherwise} \end{cases}$$

$$\gamma_p(A) = \begin{cases} \text{eventual value of } \dim_p (p^n A)\langle p\rangle & \text{if it is finite} \\ \infty & \text{otherwise} \end{cases}$$

$$\delta(A) = \begin{cases} 0 & \text{if } A \text{ has bounded exponent} \\ \infty & \text{otherwise.} \end{cases}$$

We can now recast Theorem 2 in the form promised in the Introduction.

Theorem 19. Two abelian groups A_1, A_2 are elementarily equivalent iff they have the same invariants $\alpha_{p,n}, \beta_p, \gamma_p, \delta$.

Proof: Assume first that A_1, A_2 are elementarily equivalent. Suppose a particular invariant of A_1 is finite and equal to the integer k. Then this fact is expressible by a set of first order sentences (Exercise 7), and hence the corresponding invariant of A_2 has the same value. If on the other hand a given invariant of A_1 is ∞, then the same argument shows that the corresponding invariant of A_2 is ∞. Thus A_1, A_2 have the same invariants.

We turn to the more interesting half of the theorem. Suppose that A_1, A_2 have the same invariants. Let A_1', A_2' be saturated elementary extensions of A_1, A_2. of equal cardinality. By the above, A_1', A_2' have the same invariants. It follows directly from Theorem 2 that A_1', A_2' are isomorphic, and are in particular elementarily equivalent. Hence A_1, A_2 are elementarily equivalent.

Corollary 20. Let $A_1 \subseteq A_2$ be abelian groups. Then the following are equivalent:

1. A_1 is an elementary substructure of A_2.
2. A_1 is a pure subgroup of A_2 and A_1, A_2 have the same invariants.

Proof: Apply Theorem V.5. 28 and Lemma 9.2.

Definition 21. Any group of the following form will be called a Szmielew group:

$$A = \oplus\Sigma(Z/p^n)^{(\alpha_{p,n})} \oplus\Sigma Z_p^{(\beta_p)} \oplus \Sigma(Z/p^\infty)^{(\gamma_p)} \oplus \underset{\sim}{Q}^{(\delta)} \text{ ; here}$$

$\alpha_{p,n}, \beta_p, \gamma_p$ are $\leq \aleph_0$ and δ is 0 or 1.

Corollary 22. Every group is elementarily equivalent to a Szmielew group.

Proof: Every possible combination of invariants already occurs in a Szmielew group.

§2. Construction of Saturated Groups.

We will deal with the converse to Theorem 2 and with elimination of quantifiers for the theory of abelian groups. In the statement of Theorem 2 we paid little attention to the values of the invariants $\alpha_{p,n}, \beta_p, \gamma_p, \delta$, but in the meantime we have acquired additional information (see the various lemmas preceding Definition 18). Thus we can now prove:

Theorem 23. let A be an abelian group, κ an uncountable cardinal. Then A is κ-saturated iff A can be put in the form:

$$(*)\quad A = \prod_p \left[\oplus\Sigma(Z/p^n)^{(\alpha_{p,n})} \oplus \Sigma Z_p^{(\beta_p)}\right]^{-p} \oplus \Sigma\, (Z/p^\infty)^{(\gamma_p)} \oplus \underset{\sim}{Q}^{(\delta)}.$$

where:

(**) $\alpha_{p,n}, \beta_p, \gamma_p, \delta = \alpha_{p,n}(A), \beta_p(A), \gamma_p(A), \delta(A)$ respectively if the corresponding invariant is finite; and otherwise

$\alpha_{p,n}, \beta_p, \gamma_p,$ or $\delta \geq \kappa$ (when the corresponding invariant is infinite).

Proof: If A is κ-saturated, then (*) and (**) were established in the proof of Theorem 2. At this point only the converse is interesting.

Assume therefore that A satisfies (*),(**). We must consider an arbitrary subset A_0 of A having cardinality $< \kappa$, and for any type over A_0 realized by an element b of an elementary extension B of A, we must find an element a of A realizing the same type. We may take B κ-saturated as well; our plan will then be to find an automorphism of B moving b into A. More precisely, we seek an A_0- automorphism $f: B \longrightarrow B$ such that f(b) is in A; we may then take a = f(b).

In §1 we gave a detailed analysis of the structure of B.

We now repeat this analysis, but this time we take into account the structure of A as a subgroup of B. This requires little additional effort.

Consider then a κ-saturated B elementarily extending A. Abbreviate (*) as follows:

$A = \prod_p \bar{A}_p \oplus A_d$ (A_d is the divisible part of A).

Since A_d is divisible, it is a direct summand of the divisible part B_d of B; say $B_d = A_d \oplus C_d$. When we analyze B_d as in Proposition 7, we may treat A_d, C_d separately.

We know we may write $B = B_r \oplus B_d$ with B_r Z-Hausdorff and Z-complete. B_r contains a copy of $A_r = \prod_p \bar{A}_p$ as a pure subgroup (A_r is pure in A, and A is pure in B; hence considering the injection $A_r \longrightarrow B \longrightarrow B_r$ we get the desired copy of A_r, which we may identify with A_r).

Write $B_r = \prod_p \bar{B}_p$ as in Proposition 6 (our notation is momentarily peculiar in that we have not yet defined the subgroup B_p of $\bar{B}_p$). A_p is a subgroup of $\bar{B}_p$, and the proof of Proposition 10 shows that A_p may be extended to a sum of cyclic Z_p-modules dense in $\bar{B}_p$; in other words we may take

$B_p = A_p \oplus C_p$ where C_p is a direct sum of cyclic Z_p-modules Z/p^n, Z_p.

Summarizing the above:

$B = \prod_p (\bar{A}_p \oplus \bar{C}_p) \oplus A_d \oplus C_d$. Thus A sits inside B in the simplest possible fashion. Recall our objective: we now want to move b inside A while holding A_o fixed. Visualize A_p, C_p, A_d, C_d laid out in some fashion as direct sums of the basic abelian groups. Consider an invariant $\alpha_{p,n}, \beta_p, \gamma_p$, or δ and the corresponding component X of C_p or C_d. A and B have the same invariants. Therefore:

If the given invariant is finite then $X = 0$.

By assumption:

If the given invariant is ∞ then in A there are at least κ copies of the corresponding basic abelian group.

It should now be clear, or almost clear, that b can be carried into A "componentwise" by an automorphism fixing A_o. We will say a little more:

Let $b = a + c$ with a in A, c in $C = \prod_p \bar{C}_p \oplus C_d$. Without loss of generality take $b = c$. We may assume that A_o is contained in $\prod_p A_p \oplus A_d$, and that c is in $\prod_p C_p \oplus C_d$ (otherwise replace A_o by a suitable set A_1 whose closure contains A_o, and replace c similarly by a countable subset of $\prod_p C_p \oplus C_d$). By our previous remarks there is adequate room to move the various components of c (or of fewer than κ c's) into A without disturbing A_o.

We now return to a topic raised fleetingly in the first lines of §1- the expressive capacity of the first order language of abelian groups. Let $a_1,\ldots,a_n$ be elements of an Abelian group A. It turns out that there are only two kinds of statements one can make about the way $\bar{a}$ sits in A:

1. Statements about A (from the present point of view this is a degenerate case).
2. Divisibility statements: "n divides $\Sigma m_i a_i$" (taking $n = 0$, this includes torsion statements- admittedly this is cheating).

Obviously it follows from §1 that the only statements one can make about A concern the invariants $\alpha_{p,n}, \beta_p, \gamma_p, \delta$. We left the precise version of this remark to the reader as Exercise 7.

The theorem we wish to prove is the following:

<u>Theorem 24</u>. Let $F(x_1,\ldots,x_n)$ be a formula in the language of abelian groups. Then F is equivalent to a Boolean combination of

formulas of the form:

1. sentences (no mention of $x_1,\dots,x_n$)
2. divisibility statements ("n divides $\Sigma m_i x_i$")

We will actually prove Theorem 24 in a different formulation:

Theorem 25. Adjoin constants $c_1,\dots,c_n$ to the language of abelian groups and let T be a complete theory in the extended language. Let T_0 be the subtheory of A consisting of formulas in T of the form 24.1,2 (with x_i replaced by c_i), or their negations. Then T_0 proves T.

The equivalence between Theorem 24 and Theorem 25 is the logician's answer to "abstract nonsense".

Theorem 25 proves Theorem 24:

Let $F(x_1,\dots,x_n)$ be a formula in the language of abelian groups. Adjoining constants $c_1,\dots,c_n$ to the language, we replace $F(x_1,\dots,x_n)$ by the sentence $F(c_1,\dots,c_n)$. Call a sentence S in the extended language a reason for F iff:

1. S is a Boolean combination of sentences of forms 24.1,2
2. S proves F (relative to the theory of abelian groups, of course).

Let T_1 be the theory:

$\{F(\bar{c})\} \cup \{\neg S(\bar{c}): S \text{ a reason for } F\}$.

It is a consequence of Theorem 25 that T_1 is inconsistent. If T_1 were consistent it would have a consistent complete extension T to which Theorem 25 applies. Then some finite subset S of T_0 would prove $F(\bar{c})$; replacing S by the conjunction of all statements in S (which we again denote S), we now have a reason S for F which is consistent with T_1 - but this is nonsense.

So T_1 is inconsistent, and hence some finite subset $\{F(\bar{c})\} \cup \{-S_i: i = 1,\dots,k\}$ is inconsistent. High school logic shows then that F is equivalent to $S_1 \vee \ldots \vee S_k$, proving Theorem 24.$\dashv$

In order to prove Theorem 25 we have to make a reasonably thorough study of the ways in which finitely generated abelian groups can sit inside general abelian groups A- however our methods allow us to restrict our attention to saturated A, which is a comfort.

The key to the proof is the use of relative pure-injective hulls, a variation on the motif of Chapter V §5. Since nothing is gained for the moment by restricting ourselves to the context of abelian groups, we describe this idea ih the context of modules over a ring R.

Definition 26. Let $M_1 \subseteq M_2$ be modules with M_2 pure-injective.

1. A formula $F(x)$ with one free variable x is called positive primitive iff F has the following form:

"$\exists \bar{y} \;\; \bar{y}A = x\bar{r}+\bar{m}$" ; here A is a matrix, the r_i lie in R, and the m_i lie in M_2 (usually in M_1).

2. A set S of positive primitive formulas defined over M_1 is consistent iff it is satisfied in M_2.

3. S is a (pos. prim.) type iff S is maximal among consistent sets of positive primitive formulas.

(Notice that our definition of consistency is reasonable only because M_2 is pure-injective. In particular types exist.)

4. A module $\bar{M}$ is a relative pure-injective hull of M_1 in M_2 iff:

i. $\bar{M}$ is a pure-injective pure submodule of M_2 containing M_1.

ii. $\bar{M}$ is prime among structures satisfying i, i.e. M can be embedded in any other M' satisfying (i) in such a way that

$$\begin{array}{ccc} \bar{M} & \xrightarrow{f} & M' \\ & \nwarrow \;\; \nearrow & \\ & M_1 & \end{array}$$

commutes and f is a pure map.

Theorem 27. If $M_1 \subseteq M_2$ and M_2 is pure-injective then M_1 has a relative pure-injective hull $\overline{M}$ in M_2.

Proof: We make use of types in the sense of Definition 26. If every type over M_1 which is realized in M_2 is already realized in M_1, then it follows that M_1 is already pure in M_2 and pure-injective, and there is nothing to prove. In the contrary case we extend M_1 by adjoining realizations of various types realized in M_2. More precisely, we define a sequence M_α of submodules of M_2 by transfinite induction, starting with M_1:

1. $M_{\alpha+1}$ is obtained by adjoining a realization a of a type S to M_α. (If every type S over M_α is already realized in M_α then just take $M_{\alpha+1} = M_\alpha$.)

2. For δ a limit ordinal, $M_\delta = \bigcup_{\alpha<\delta} M_\alpha$.

Let $\overline{M} = \bigcup_\alpha M_\alpha$; then $\overline{M}$ is at least a submodule of M_2 containing M_1. Furthermore by the construction $\overline{M}$ realizes all types defined over $\overline{M}$. Hence $\overline{M}$ is pure in M_2 and pure-injective. The point of the construction appears in verifying condition 26.4.ii.

Suppose then that M' is a second pure-injective pure submodule of M_2 containing M_1. We must find a pure embedding f of $\overline{M}$ into M' over M_1. Naturally we construct f by transfinite induction. In other words we seek a sequence f_α of partial embeddings of $\overline{M}$ into M' satisfying the following:

1. $f_\alpha : M_\alpha \longrightarrow M'$

2. f_α is type-preserving, i.e. whenever $\exists \bar{y}\ \bar{y}A = a\bar{r}+\bar{m}$, then $\exists \bar{y}\ \bar{y}A = f(a)\bar{r}+\overline{f(m)}$.

(Notice that the type of a in M_2 coincides with its type in $\overline{M}$, and the type of $f(a)$ in M_2 coincides with its type in M'.)

We may evidently take f_1 as the identity map on M_1, and we will take unions at limit ordinals ($f_\delta = \bigcup_{\alpha<\delta} f_\alpha$). To obtain

$f_{\alpha+1}$, recall that $M_{\alpha+1} = \langle M_\alpha, a\rangle$ for some a realizing a type S over M_α. Let fS be the corresponding type over $f_\alpha[M_\alpha]$.

Claim A. fS is realized in M' by an element b.

Claim B. We may define a type-preserving extension of f by

$$f_{\alpha+1}(m+ar) = f_\alpha(m) + br.$$

The verification of these claims involves a routine unraveling of the definition 26.3, which we omit.

We return to the context of abelian groups.

Proof of Theorem 25:

Let T, T_0 be as in the statement of Theorem 25, i.e. T is a complete theory of abelian groups A with distinguished elements $c_1,\ldots,c_n$, and T_0 is the fragment of T consisting of:

1. a complete theory of abelian groups A
2. divisibility statements concerning elements of $\langle c_1,\ldots,c_n\rangle$.

We must prove that T_0 is already complete, and hence proves T. Consider therefore two models A_1, A_2 of T_0. We must show that $\langle A_1,\bar{c}\rangle$ and $\langle A_2,\bar{c}\rangle$ are elementarily equivalent. Without loss of generality we may take A_1, A_2 saturated and of equal cardinality. By (1) above, A_1 and A_2 are isomorphic; we must prove that A_1 and A_2 are even isomorphic over $\langle\bar{c}\rangle$ (by (2) the group $C=\langle\bar{c}\rangle_1$ generated by $\bar{c}$ in A_1 is isomorphic with the corresponding group $\langle\bar{c}\rangle_2$ generated in A_2.)

Loosely speaking, we must prove that T_0 describes the manner in which C sits in A_1 or A_2. More precisely:

Claim. Let $\bar{C}$ be a relative injective hull of C in A_1. Then there is a pure embedding of $\bar{C}$ in A_2.

(Verification: Consider the diagram

$$\begin{array}{ccccc} & \nearrow & A_1 & \xrightarrow{f} & \\ C & & & & A_1 \oplus_C A_2. \\ & \searrow & A_2 & \xrightarrow[g]{} & \end{array}$$

It is trivial to prove that f,g are pure maps (the divisibility statements- and negations thereof- as in (2) are just what is needed for this). It follows that $\overline{C}$ is a pure-injective hull of C in $A_1 \oplus_C A_2$ and that A_2 is a pure pure-injective submodule of $A_1 \oplus_C A_2$. Thus $\overline{C}$ admits a pure embedding into A_2.)

Using the Claim we finish the argument as follows. Since $\overline{C}$ is pure-injective and pure in A_1, A_2 we may write:

$A_1 = \overline{C} \oplus B_1$.

$A_2 = \overline{C} \oplus B_2$.

We note that the cardinality of $\overline{C}$ is at most $2^{\aleph_0}$, and that there is no loss of generality in assuming that the cardinalities of A_1, A_2 are greater than $2^{\aleph_0}$. Since the invariants $\alpha_{p,n}, \beta_p, \gamma_p, \delta$ are additive it follows that:

B_1, B_2 have the same invariants and furthermore

B_1, B_2 are saturated (Theorem 23).

Hence B_1, B_2 are isomorphic, and therefore A_1 is isomorphic with A_2 over C (even over $\overline{C}$), as desired.

§3. Notes.

The decidability of the theory of abelian groups and a strong version of Theorem 25 were first proved in |52| by direct syntactical methods. The source for our treatment via saturated models is |20|. Syntactical methods yield results which are more precise and explicit in the context of abelian groups, but which generalize less readily to general modules (see Exercise 11).

The main topic dealt with in this chapter was the generalization of the structure theorem for abelian groups to the context of saturated groups. Model theory offers other views of structure theory which are of interest in the context of abelian groups. For the use of infinitary logic to generalize Ulm's Theorem see |18|, and for the application of techniques growing out of stability theory see |47|.

Exercises.

§1.

1. Let K be a class of abelian groups and let A be an abelian group. Show that there is a subgroup B of A maximal subject to: B is a direct sum of elements of K.
2. Let p be a prime and let D_p be a divisible abelian group in which every element is annihilated by a power of p. Show that D_p contains a copy of Z/p^{∞}.
3. Generalize Lemma 9 to modules over principal ideal domains.
4. (Proposition 12.) Let A be an abelian $\aleph_1$-saturated group, and suppose A does not have finite exponent. Prove in succession: $\delta \geq 1$, $\delta \geq \kappa$.
5. Complete the proof of Lemma 16.
6. Prove Lemma 17. The two cases in Lemma 17 correspond to the two cases in Lemma 16.
7. Formalize the following by formulas in the quantifier complexity class E_2:
   i. "$\dim_p (p^{n-1}A)\langle p\rangle/(p^nA)\langle p\rangle \geq k$"
   ii. "$\dim_p (p^nA/p^{n+1}A) \geq k$"
   iii. "$\dim_p (p^nA)\langle p\rangle \geq k$"
   iv. "$nA \neq 0$"
8. Compute the invariants of $\langle Z,+\rangle$ and relate this to the structure of nonstandard models of arithmetic.
9. Prove: the theory of abelian groups is decidable (i.e. there

is a mechanical procedure for checking whether a putative theorem about abelian groups is correct). There is a useful bit of abstract nonsense following Lemma 25 which is relevant here. The main ingredient is Theorem 19.

§2.

10. Generalize the material of this chapter to modules over principal ideal domains (or over Dedekind domains). Consult |27| for algebraic ingredients.

11. Let T be the theory of modules over a given ring R (if R is uncountable then T is uncountable). Show that every model of T has a saturated elementary extension (no false axioms of set theory- like the continuum hypothesis- are permitted in this exercise). More explicitly, generalize Theorem 23 as far as possible to modules, using relative pure-injective hulls. (This result- that any theory of modules is stable- is due to Fisher |23|.)

VII. $\aleph_1$-categorical Fields

Introduction.

Much of the material of Chapters II-V has its roots, historically and psychologically, in the Hilbert Nullstellensatz as treated in Chapter I§2. All proofs of the Nullstellensatz have this much in common: they rely in one way or another on something like Steinitz' classification of simple field extensions $K(\alpha)$ as:

1. algebraic over K (classified by the minimal polynomial of α over K)

or 2. transcendental over K (unique up to K-isomorphism).

(This is apparent if Noether's Normalization Theorem is used but rather well hidden in the place extension theorem.)

In §1 we will give a proof of the Nullstellensatz which does not use the structure theory for fields, but does use the fact that there <u>is</u> such a structure theory.

<u>Definition</u>. A theory T is <u>$\aleph_1$-categorical</u> iff any two models M_1, M_2 of T having cardinality $\aleph_1$ are isomorphic.

Let T be the theory of algebraically closed fields of fixed characteristic p (a prime or zero). Then T is $\aleph_1$-categorical. Reflecting on the usual proof of this fact, one sees that Steinitz' classification- or more accurately, the <u>existence</u> of some classification like Steinitz'- is implicit in the $\aleph_1$-categoricity of T.

Model theorists have devoted a great deal of attention in recent times to $\aleph_1$-categorical theories, and more generally to theories whose models possess a "structure theory" in various senses. In §2 we will outline some of the known facts, which are applied in §3 to prove:

<u>Theorem</u>. Let T be a complete theory of fields. Suppose T is $\aleph_1$-categorical. Then T is one of the following:

1. the theory of a particular finite field
2. the theory of all algebraically closed fields of some fixed characteristic.

§1. <u>The Nullstellensatz Revisited</u>.

<u>Convention</u>. Throughout this section T will be the theory of algebraically closed fields of some fixed characteristic.

The following is a version of the Nullstellensatz.

<u>Theorem 1</u>. (Hilbert Nullstellensatz). Let K be an algebraically closed field, K' an extension of K, and let V be the variety determined by polynomials $p_1,\ldots,p_k \in K[x_1,\ldots,x_n]$. If $V_{K'}\neq\emptyset$ then $V_K\neq\emptyset$ (i.e. if V has a point in $(K')^n$ then V already has a point in K^n).

From the point of view of Chapter I §2 this is half of the Nullstellensatz- the half amenable to model theoretic methods. To obtain the full Theorem I.2 it is necessary to repeat the algebraic half of the argument given there (or use Chapter III §6). Our current proof of the Nullstellensatz will use two facts about fields and one theorem of logic.

<u>FACT A</u>. T is <u>inductive</u> (i.e. the union of an increasing chain of models of T is again a model of T).

<u>FACT B</u>. T is <u>$\aleph_1$-categorical</u> (i.e. any two models of T of cardinality $\aleph_1$ are isomorphic).

<u>FACT C</u>. Let P be a theory of pairs of structures $A_0 \subseteq A_1$ (e.g. rings with a two-step filtration). Let κ be an infinite cardinal. Suppose there is a model $A = (A_0, A_1)$ of P for which A_0 is infinite. Then there is a model $B = (B_0, B_1)$ of P in which B_0 has cardinality κ.

Notice that Fact A is perfectly harmless. The bulk of the axiom systems one encounters in daily life are inductive, as is usually clear upon inspection of the axioms. As we said in the introduction, Fact B is roughly equivalent to the existence of a definitive structure theory for field extensions. Fact C will be recognized as a minor variant of the Löwenheim-Skolem theorem (cf. Chapter 0 § 5), adapted to our present purposes.

Proof of the Nullstellensatz:

Call a model K of T variety complete iff K satisfies the Hilbert Nullstellensatz (with respect to any extension K', and any polynomials $p_1,\dots,p_k$). We make the following claims:

Claim A. There is a variety complete K of cardinality $\aleph_1$.

Claim B. All models K of T having cardinality $\aleph_1$ are variety complete.

Claim C. All models K of T are variety complete.

Since our final claim is exactly the Hilbert Nullstellensatz, it suffices to verify our claims.

Verification A: We prove Claim A using Fact A alone. We did this previously in Chapter III §1: starting with any model K of T having cardinality $\aleph_1$, we extend K repeatedly until it becomes variety complete. The inductivity of T allows us to continue this process indefinitely, taking unions at limit ordinals. See Exercise 1.

Verification B: Given Claim A and Fact B, Claim B is immediate (since T has essentially only one model of cardinality $\aleph_1$).

Verification C: Given Claim B and Fact C, Claim C is immediate. In more detail:

If there is a counterexample to Claim C, it consists of a pair of fields (K,K') and some polynomials $p_1,\dots,p_k$ satisfying the following conditions:

1. K is algebraically closed.

2. $p_1,\ldots,p_k$ have coefficients in K.

3. $V_K(p_1,\ldots,p_k) = \emptyset$.

4. $V_{K'}(p_1,\ldots,p_k) \neq \emptyset$.

By Fact C, given any such (K,K') we can find another pair (K_1,K_1') with the same properties, in which K_1 has cardinality $\aleph_1$. But this would contradict Claim B.

Using the same style of proof one can show that models of $\aleph_1$-categorical theories have all nice properties that could reasonably be expected (although the subtler results in this vein rely on substantial refinements of Fact C- see [ ]).

§2. Introduction to Categoricity.

In this section we will summarize some of the known facts concerning categoricity, while proving little or nothing. This material will be used in §3 to study theories of fields.

Definition 2. Let κ be an infinite cardinal. A theory T is κ-categorical iff all models of T of cardinality κ are isomorphic.

Example 3.

1. The theory T of vector spaces over a fixed finite field F_q is κ-categorical for all infinite κ.

2. The theory T of vector spaces over the rational field, or equivalently the theory of torsion-free divisible abelian groups, is κ-categorical for all uncountable κ.

3. The theory T of dense linear orderings without end-points is $\aleph_0$-categorical (Exercise 6) but is not κ-categorical for any uncountable κ.

Notice that there is a structure theory for structures of types 1,2 but not for structures of type 3 (there are in fact 2^κ nonisomorphic dense linear orderings of any uncountable cardinality κ).

<u>Theorem 4</u>. Suppose T is categorical in one uncountable cardinali-
ty. Then T is categorical in all uncountable cardinalities. In
other words there are only two types of categoricity: $\aleph_0$-categori-
city and $\aleph_1$-categoricity.

We are going to neglect $\aleph_0$-categoricity completely in favor
of $\aleph_1$-categoricity (the interested reader may consult | , |).

One of the most useful consequences of $\aleph_1$-categoricity is a
weak version of the Steinitz structure theory for fields. In the
context of fields the following is obvious:
<u>Obvious Fact</u>. Let T be the theory of algebraically closed fields.
Let F be a countable substructure of a model K of T. Let
$S_K(F)$ be the set of 1-types realized in K over F (Chapter 0 §2).
Then $S_K(F)$ is countable.

One way to verify this is as follows. Suppose a,b are
elements of K. If F(a),F(b) are <u>isomorphic</u> over F via an
isomorphism taking a to b, then this isomorphism extends to an
automorphism of K, and it follows that a,b have the <u>same type</u>
over F. On the other hand there are only countably many <u>F-iso-</u>
<u>morphism types</u> of extensions F(a), so there are only countably
many <u>types</u> over F.

We contrast this fact with something even more obvious:
<u>Obvious Fact</u>. Let T be the theory of dense linear orderings
without endpoints. Then the reals $\underset{\sim}{R}$ form a model of T, and $\underset{\sim}{R}$
contains a countable substructure, namely the rationals $\underset{\sim}{Q}$, such
that $S_{\underset{\sim}{R}}(Q)$ is uncountable.

In fact since any two elements of $\underset{\sim}{R}$ can be separated by
a rational, no two points of $\underset{\sim}{R}$ realize the same type over Q.

The following definition is of the greatest importance.
<u>Definition 5</u>. A theory T is $\aleph_0$-<u>stable</u> iff for any model M of
T and any countable substructure F of M, the set $S_M(F)$ of

types of elements of M over F is countable.

Notice that $\aleph_0$-stability already hints at the existence of a structure theory for simple extensions. As we shall see, $\aleph_0$-stability lends itself to direct application in algebraic contexts. Unlike $\aleph_1$-categoricity, $\aleph_0$-stability is preserved by a variety of algebraic constructions.

Example. Let T be the theory of abelian groups of fixed exponent, say 120. T is $\aleph_0$-stable, because there is an adequate structure theory for such groups. (Exercise: verify this.)

Let A be a given algebraic structure. We will call A $\aleph_0$-stable if the theory Th(A) is $\aleph_0$-stable. Referring back to the definition of $\aleph_0$-stability, this means that we must consider all models A' elementarily equivalent to A, all countable. substructures F of A', and compute the number of types realized in A' over F. For this purpose it is sensible to consider only the $\aleph_1$-saturated models A', since they realize the maximum number of types.

The connection between categoricity and stability is quite simple:

Theorem 6. Let T be $\aleph_1$-categorical. Then T is $\aleph_0$-stable.

Proof: The argument resembles the proof given in §1 for the Nullstellensatz. The steps are as follows:

Claim A. There is a model M of T having cardinality $\aleph_1$ such that for any countable substructure $F \subseteq M$, $S_M(F)$ is countable.

Claim B. All models M of T having cardinality $\aleph_1$ are as in Claim A.

Claim C. All models M of T are as in Claim A.

Claim A in fact holds for any theory T whatsoever, but the proof is delicate and we will not give it. Given Claim A, Claim B follows at once, and Claim C soon thereafter.

In concrete algebraic contexts one can often deduce consequences of $\aleph_0$-stability directly. However one often finds the following results more pliable than the definitions:

Theorem 7. Let M be a model of a theory T. If either of the following conditions is satisfied then T is not $\aleph_0$-stable:

1. There is a definable binary relation R which linearly orders an infinite subset S of M.
2. There is an n-ary relation R and an infinite subset S of M such that for any $\bar{a} = (a_1, \ldots, a_n)$ in S there are permutations $\bar{a}^\sigma, \bar{a}^\tau$ of $\bar{a}$ such that:
   i. $R\bar{a}^\sigma$
   ii. $\neg R\bar{a}^\tau$.

Notice that 1 is a special case of 2. We will use 2 in all its glory in §3. A relation R satisfying 2.i,ii is called:

i. connected
ii. asymmetric.

A typical consequence of Theorem 7 in the context of $\aleph_0$-stable rings is the descending chain condition on left or right ideals (since right or left divisibility does not linearly order any infinite set).

Condition 2 of Theorem 7 is called "Ehrenfeucht's Condition".

We mention one more general fact concerning $\aleph_0$-stable theories.

Theorem 8. Let T be an $\aleph_0$-stable theory, let M be a model of T, and let F be a countable substructure of M. Let $S^n_M(F)$ be the set of n-types realized over F in M. Then $S^n_M(F)$ is countable.

We will not prove this, but we remark that for $n = 1$ the statement to be proved coincides with the definition of $\aleph_0$-stability, and that the general case follows easily by induction.

Corollary 9. Let R be a ring, A an algebra of finite type over R, and suppose that Th(R) is $\aleph_0$-stable. Then Th(A) is

$\aleph_0$-stable.

The corollary follows quickly but not instantaneously from Theorem 8 (Exercise 7).

§3. $\aleph_1$-categorical fields.

The main theorem of this section will be:

Theorem 10. Let T be a complete $\aleph_1$-categorical theory of fields. Then T is either:

1. the theory of a particular finite field

or 2. the theory of all algebraically closed fields of some fixed characteristic.

In fact one proves the following:

Theorem 11. Let T be a complete $\aleph_0$-stable theory of fields. Then T is one of the above.

We hasten to add that $\aleph_1$-categoricity is usually far stronger than $\aleph_0$-stability, and theories of fields are in this respect a degenerate case.

The proof of Theorem 11 involves a number of ad hoc arguments coupled with two facts of more general significance:

Theorem 12. If F is a field with $\aleph_0$-stable theory and K is a finite dimensional extension of F then K is $\aleph_0$-stable. (The converse is false: take F = the reals, K = the complexes, and notice that F has a definable linear ordering- cf. Theorem 7.)

Fact 13. The theory of cyclic extension of fields (i.e. Galois extensions with cyclic Galois group) has been completely worked out, modulo some problems concerning roots of 1 which we will be able to avoid.

Theorem 12 is a special case of Corollary 9. As far as Fact 13 is concerned, we will need the following:

Theorem 14. Let K be a cyclic extension of the field F, of degree n, and suppose the Galois group of K over F is generated by the element σ.

1. If n is not divisible by the characteristic of F and if x^n-1 splits in F then K is generated by n^{th} roots of elements of F.
2. If $n = char(F)$ then $K = F[\alpha]$ for some root α of an irreducible equation $x^n - x - a = 0$, with a in F.
3. Let $Tr: K \longrightarrow F$ be the trace map from K to F. Then $ker(Tr) = Im(\sigma-1)$; explicitly, if x is in K and $Tr(x) = 0$ then for some y in K, $x = \sigma y - y$.

We are primarily interested in the first two points, with the additional assumption that n is prime. The third point will be needed just at the end of this section (F will be the Galois field F_p and K will be a finite field).

For proofs of these statements consult |2|.

We now have an adequate logical and algebraic background to begin the proof of Theorem 11. We will establish two facts by direct argument:

<u>Theorem 15</u>. Let F be an infinite field with $\aleph_0$-stable theory, $a \in F$, p a prime. Then:

1. $a^{1/p} \in F$.
2. If $char(F) = p$ then x^p-x-a has a root in F.

Before proving Theorem 15, let us see that we have in fact collected the ingredients for a proof of Theorem 11.

<u>Proof of Theorem 11</u>:

Suppose there is an infinite field F_0 with $\aleph_0$-stable theory which is not algebraically closed. Let p be minimal such that:

(*) there is a field F containing F_0 such that F has $\aleph_0$-stable theory and F has an extension of degree p.

By Theorem 12 and elementary Galois theory, p is a prime.

Fix F as in (*). Then x^p-1 splits over F (since otherwise p is not minimal). If $p \neq \text{char}(F)$, then by Theorem 14.1 there is an irreducible polynomial of the form x^p-a with a in F, contradicting Theorem 15.1. Thus $p = \text{char}(F)$, and by Theorem 14.2 there is an irreducible polynomial x^p-x-a with a in F, contradicting Theorem 15.2.

The rest of this section is devoted to a proof of Theorem 15. Before treating Theorem 15.1, we insert a lemma.

Lemma 16. Let A be an abelian group whose theory is $\aleph_0$-stable. Then $A = B \oplus D$ where D is divisible and B has finite exponent.

Proof: By Chapter VI §1 A is elementarily equivalent to a group of the form $R \oplus D$ with D divisible and R complete Hausdorff in the Z-topology. We claim that R is of finite exponent, in other words that R is discrete in the Z-topology.

In the contrary case choose a countable nondiscrete subgroup $H \subseteq R$. Then the closure $\bar{H}$ of H in R is uncountable (this is true for any nondiscrete complete Hausdorff topological group), and every point in $\bar{H}$ realizes a different type over H (even in the language of abelian groups). This contradicts the $\aleph_0$-stability of $\text{Th}(A)$.

Thus A is elementarily equivalent to $R \oplus D$ with D divisible and R of finite exponent n. It follows that nA is divisible, and hence $A = B \oplus D$ with B of exponent n and $D = nA$ divisible. This proves the lemma.

Proof of Theorem 15.1:

Let A be the multiplicative group of nonzero elements of the infinite $\aleph_0$-stable field F. Then A also has $\aleph_0$-stable theory (Exercise 10). Thus $A = B \times D$ for some B of exponent $n < \infty$ and divisible D (we will use multiplicative notation in dealing with A). Since $x^n = 1$ for x in B, B must be finite and hence cyclic (since F is a field). Let $B = \langle a \rangle$. We claim $a = 1$.

Suppose on the contrary that $a \neq 1$, so that $n > 1$. We will apply Theorem 7.2. Define a relation R on n-tuples of F by:

$$Rx_0 \ldots x_{n-1} \text{ iff } x_0 + ax_1 + \ldots + a^{n-1}x_{n-1} \in (F^x)^n.$$

Define $f(\bar{x}) = \sum x_i a^i$, so that

$R\bar{x}$ iff $f(\bar{x}) \in (F^x)^n$. Notice that a cyclic shift of $x_0, \ldots, x_{n-1}$ multiplies $f(\bar{x})$ by the corresponding power of a, so that if $f(\bar{x}) \neq 0$ then there is exactly <u>one</u> cyclic permutation σ of $\bar{x}$ such that $f(\bar{x}^\sigma) \in (F^x)^n$. Hence R is clearly connected and asymmetric (in the sense of Theorem 7.2) except possibly for n-tuples $\bar{x}$ such that $f(\bar{x}) = 0$. In the latter case let σ be the transposition (01). Then:

$$f(\bar{x}^\sigma) = (a-1)(x_0 - x_1) + f(\bar{x}) \neq f(\bar{x}) = 0;$$

thus $f(\bar{x}^\sigma) \neq 0$ and our previous argument using cyclic permutations proves that R is connected and asymmetric.

This contradicts Theorem 7. It follows that $a = 1$ and $A = D$ is divisible.

We come now to the second half of Theorem 15. We claim that $\tau[F] = F$ where $\tau(x) = x^p - x$, $p = \text{char}(F)$. Notice that τ and its iterates τ^n are endomorphisms of the additive group $\langle F,+\rangle$. We have:

(*) $F \supseteq \tau[F] \supseteq \tau^2[F] \supseteq \ldots$

<u>Lemma 17</u> For some n $\tau^n[F] = \tau^{n+1}[F] = \ldots$.

<u>Proof</u>: Suppose on the contrary that the descending sequence (*) never stabilizes. Consider the collection C of all cosets $a + \tau^n[F]$ of the various $\tau^n[F]$ in $\langle F,+\rangle$; C is partially ordered by $\supseteq$. Under this partial ordering C forms a tree, and every point of C has two incomparable successors. We can therefore

select a <u>countable</u> subtree $C' = \{a_i+\tau^{n_i}[F]\}$ having $2^{\aleph_0}$ paths p. A path p through C or C' is just a decreasing sequence of cosets. If F is taken $\aleph_1$-saturated, then corresponding to each path p of C' there is at least one point x_p such that $x_p \in \bigcap p$.

The $2^{\aleph_0}$ elements x_p have different types with respect to the parameters $\{a_i\}$, and this contradicts the $\aleph_0$-stability of Th(F).

Our aim of course is to show that we can take $\hat{n} = 0$ in the previous lemma. We insert the following lemma to illustrate a certain method. Recall that the theory of vector spaces over a given finite field K is categorical in all infinite powers, and is in particular $\aleph_0$-stable.

<u>Lemma 18</u>. Let T be the theory of vector spaces V over the fixed finite field K, equipped with a nondegenerate symmetric bilinear form B("nondegenerate" means $V^{\perp} = (0)$). Then T is not $\aleph_0$-stable.

<u>Proof</u>: Let V be an infinite $\aleph_1$-saturated model of T, and let $\{a_n : n \in \underset{\sim}{N}\}$ be $\aleph_0$ linearly independent vectors in V. For any function $f : \underset{\sim}{N} \longrightarrow K$ let

p_f = "$B(x,a_n) = f(n)$" (all n).

Then each of the $2^{\aleph_0}$ types p_f is realized in V over $\{a_n\}$, and hence T is not $\aleph_0$-stable.

The relevance of this lemma arises from the following:

<u>Fact 19</u>. Let K,F be fields, with K a finite dimensional separable extension of F. Then the trace function

$Tr(xy) : K \times K \longrightarrow F$

is bilinear and nondegenerate [2].

We will use this fact in conjunction with Theorem 14.3.

Proof of Theorem 15.2:

We assume F is infinite, $\aleph_0$-stable of characteristic p, and $\tau(x) = x^p-x$. We claim $\tau[F] = F$. Let A be the field of algebraic numbers in F. We claim:

Claim A. A is infinite.

Claim B. $\tau[A] = A$.

Claim C. $\tau[F] = F$.

The main idea occurs in the proof of Claim B.

Verification A:

We know that the multiplicative group of nonzero elements of A is divisible (Theorem 15.1). Hence A is infinite or $A = F_2$. In the latter case adjoin a root i of $x^2+x+1 = 0$ to F. Then $F[i]$ is still $\aleph_0$-stable by Theorem 12. Let B be the field of algebraic elements of $F[i]$. Then $B \neq F_2$, so B is infinite. By the argument given below Claims B and C follow for $F[i]$, and the proof of Theorem 11 given above then applies to show that $F[i]$ is algebraically closed. (This already contradicts the Artin-Schreier Theorem, but we prefer to give a more explicit argument:)

In particular $F[i]$ contains a root r of $x^2+x+i = 0$. Writing $r = a+bi$ with a,b in F we compute:

$$a^2+b^2(i+1)+a+bi+i = 0,$$

and looking at the coefficient of i we get $b^2+b+1 = 0$, so that x^2+x+1 has a root in A, contradicting $A = F_2$.

Hence A is infinite.

Verification B:

We remark that $\tau[A] = A \cap \tau[F]$.

Suppose $\alpha \in A-\tau[A]$. Our intention is to treat A as a vector space over F_p (infinite dimensional by Claim A) equipped with something resembling a nondegenerate bilinear form.

Choose elements $\{a_n : n \in \underset{\sim}{N}\}$ linearly independent over the prime field F_p. For any function $f: \underset{\sim}{N} \to F_p$ consider the type

(*) p_f = "$xa_n - f(n)\alpha \in \tau[F]$" (all n).

No element realizes more than one type p_f, because $\alpha \notin \tau[F]$. On the other hand we claim that all the types p_f are realized in F if F is $\aleph_1$-saturated, violating the $\aleph_0$-stability of $Th(F)$. This contradiction will prove that $A = \tau[A]$.

Consider therefore any finite subset of a type of the form (*). We must show that it is satisfied in F. Let F_0 be the subfield of A generated by α and a finite subset of the set of parameters $\{a_n\}$. F_0 is a finite field, and in particular a cyclic extension of F_p, with Galois group generated by the map σ: $x \longmapsto x^p$. By Fact 19 the trace $Tr(xy)$ from F_0 to F_p is a nondegenerate bilinear form on F_0 over F_p, so we can satisfy the conditions:

(**) "$Tr(xa_n - f(n)\alpha) = 0$"

Now we use Theorem 14.3, taking σ as above. Then $\sigma - 1$ is just τ, so in our context Theorem 14.3 says:

$$\ker(Tr) = \operatorname{im}(\tau|_{F_0}).$$

In particular the satisfiability of (**) implies the satisfiability of (*). This completes the argument.

Verification C:

We now know that $\tau[A] = A$. We will show that $F \subseteq \tau[F]$. Fix x in F. By Lemma 17, for some n $\tau^n[F] = \tau^{n+1}[F]$. Write $\tau^n(x) = \tau^{n+1}(y)$ for a suitable y. Then $\tau^n(x - \tau(y)) = 0$. In particular $x - \tau(y)$ is algebraic, so $x - \tau(y) \in A = \tau[A]$. Write $x - \tau(y) = \tau(z)$ with z in A. Then $x = \tau(y+z)$, so $x \in \tau[F]$, as desired.

This completes the proof of Theorem 11, and hence of Theorem 10.

Theorem 10 can be extended mutatis mutandis to various classes of rings. For example:

Theorem 20. An $\aleph_1$-categorical division ring is commutative, hence finite or algebraically closed.

We will not give the proof of Theorem 20 here. The algebraic ingredients are limited, but the proof relies heavily on a purely model theoretic result |9|. We are not in a position to state the result needed, but we will cite one algebraic consequence of it, from which Theorem 20 follows.

Fact 21. Let R be an $\aleph_1$-categorical ring, F an infinite definable subfield of R. Then R is finite-dimensional over F.

§4. Notes.

We treated the Nullstellensatz in §1 as a special case of Lindstrøm's Theorem (Exercise 2). Morley made extensive use of $\aleph_0$-stability (among other things) to prove Theorem 4 in |35|. The material in §3 is due to Macintyre |58|, with the exception of Theorem 20, which is Shelah's.

There are numerous indications that variants of $\aleph_0$-stability are intimately connected with the existence or nonexistence of a structure theory for a given class of structures. A few decades ago algebraists were obsessed by chain conditions, which are heavily set theoretic in character. In many cases it has become fashionable to study the consequences of weaker conditions involving finitely generated ideals (cf. |28|); such conditions are often first order. Model theoretic conditions like $\aleph_0$-stability (and its cousins) provide large classes of rings, groups, etc. for which it is reasonable to seek a structure theory. See [49].

Exercises.

§1.

1. Fill in the details in Verification A (proof of the Nullstellensatz), checking in particular that the resulting field has

cardinality $\aleph_1$.

2. Let T be a complete inductive κ-categorical theory (κ infinite). Show that T is model-complete.

3. Exhibit a complete inductive theory T which is not model-complete.

4. Exhibit a complete inductive non-model complete theory T such that $\underset{\sim}{E}_T$ is κ-categorical for all infinite κ.

5. Suppose T is model complete. Prove that T is inductive.

§2.

6. Verify Example 3.3.

7. Prove Corollary 10 from Theorem 9.

8. Let B be a Boolean algebra with $\aleph_0$-stable theory. Show that B is finite.

9. Give an explicit construction of $2^{\aleph_1}$ nonisomorphic real closed fields of cardinality $\aleph_1$.

10. Justify the first step in the proof of Theorem 15.1.

§3.

11. Let R be an infinite ring without nilpotents such that Th(R) is $\aleph_1$-categorical. Show that R is the product of an algebraically closed field with finitely many finite fields.

12. Use Fact 21 to prove Theorem 20.

BIBLIOGRAPHY

1. Artin, E., "Über die Zerlegung definiter Funktionen in Quadrate", Hamb. Abhand. 5 (1927), 100-115.

2. ________, Algebraic Numbers and Algebraic Functions, London, 1968.

3. Artin, E. and Schreier, O., "Algebraische Konstruktion reeler Körper", Hamb. Abhand. 5 (1927), 85-99.

4. Ax, J.,"Zeroes of polynomials over finite fields", §2, Amer. J. Math. 86 (1964), 255-261.

5. _____ , "The elementary theory of finite fields", Ann. Math. 88 (1968), 239-271.

6. Ax, J. and Kochen, S., "Diophantine problems over local fields: I", Amer. J. Math. 87 (1965), 605-630.

7. ________________, "Diophantine problems over local fields: II", Amer. J. Math. 87 (1965), 631-648.

8. ________________, "Diophantine problems over local fields: III", Annals Math. 83 (1966), 437-456.

9. Baldwin, J., "α_T is finite for $\aleph_1$-categorical T," Trans. AMS 181 (1973), 37-51.

10. Belegradek, O., "Algebraically closed groups", Algebra i Logika 13 (1974), 239-255 (English trans. pp. 135-143).

11. Chang, C.-C. and Keisler,J., Model Theory, North-Holland, 1973.

12. Cherlin, G., "Model-theoretic algebra", to appear in J. Symb. Logic 40 (1975).

13. Cohn, P.M., "On the free product of associative rings: II", Math. Zeitschr ift 73 (1960), 433-456.

14. _________, "Rings with a weak algorithm", Trans. AMS 109 (1963), 332-356.

15. _________, "The Embedding of firs in skew fields", Proc. London Math. Soc. 23 (1971), 193-213.

16. _________, Free Rings and their Relations, London, Academic Press, 1971.

17. _________, "Skew fields of fractions, and the prime spectrum of a general ring", in Tulane Symposium on Rings and Operator Algebras, 1970-71, Vol. I, Lecture Notes in Mathematics, 246, Berlin, Springer-Verlag, 1972.

18. Eklof, P., "Infinitary equivalence of abelian groups", Fund. Math. 85 (1973), 305-314.

19. ________, "Categories of local functors", preprint.

20. Eklof, P. and Fisher, E., "The elementary theory of abelian groups", Ann. Math. Logic 4 (1972), 115-171.

21. Eklof, P. and Sabbagh, G., "Model-completions and modules", Annals Math. Logic 2 (1970), 251-295.

22. Ershov, Yu., "Ob elementarnoi teorii maksimal'nykh normirovannykh polei (On the elementary theory of maximal normed fields) I-II", Algebra i Logika 4 (1965), 31-69; 5 (1966) 8-40; 6 (1967) 31-37.

23. Fisher, E., "Powers of saturated modules", J. Symb. Logic 37 (1972), 777.

24. __________, "Abelian structures", preprint.

25. Hirschfeld, J. and Wheeler, W., Forcing, Arithmetic, and Division Rings, Lecture Notes in Mathematics 454, Berlin, Springer-Verlag, 1975.

26. Jacobson, N., Lectures in Abstract Algebra, Vol. III, New York, Van Nostrand, 1954.

27. Kaplansky, I., Infinite Abelian Groups, Ann Arbor, University of Michigan, 1969.

28. ____________, Commutative Rings, University of Chicago, 1974.

29. Kochen, S., "Integer valued rational functions over the p-adic numbers", in Number Theory, Proc. Sympos. Pure Math. XII, Houston, Texas, 1967, pp. 57-73, Providence, R.I., AMS, 1969.

30. __________, "The model theory of local fields", in Logic Conference, Kiel 1974 Lecture Notes in Mathematics, 499, Berlin, Springer-Verlag, 1975.

31. Lang, S., "On quasi algebraic closure", Annals Math. 55 (1952), 373-390.

32. ________, "Some theorems and conjectures in diophantine equations", Bull. AMS 66 (1960), 240-249.

33. Macintyre, A., "The word problem for division rings", J. Symb. Logic 38 (1973), 428-436.

34. ____________, "On algebraically closed division rings", preprint.

35. Morley, M., "Categoricity in power", Trans. AMS 114 (1965), 514-538.

36. Robinson, A., Introduction to Model Theory and to the Metamathematics of Algebra, Amsterdam, North-Holland, 1965.

37. ____________, "Problems and methods of model theory", in Aspects of Math. Logic (CIME Varenna, 1968), 181-266. Rome, Edizioni Cremonese, 1969.

38. ___________, "Infinite forcing in model theory", in Proceedings of the Second Scandanvian Logic Symposium, Oslo, 1970, Amsterdam, North-Holland, 1971.

39. Rogers, H., Theory of Recursive Functions and Effective Computability, New York, McGraw-Hill, 1967.

40. Roquette, P., "Bemerkungen zur Theorie der formal p-adischen Körper", Beiträge Z. Alg. u. Geometrie 1 (1971), 177-193.

41. Sabbagh, G., "Sous-modules purs, existentiellement clos et élémentaires", C.R. Acad. Sci. Paris 272 (1971) Series A, 1289-1292.

42. Sacks, G., Higher Recursion Theory, Mimeographed notes.

43. ________, Saturated Model Theory, Reading, Mass., Benjamin, 1972.

44. Schilling, O., The Theory of Valuations, New York, AMS, 1950.

45. ___________, "Die Untergruppen der freien Gruppen", Hamb. Univ. Math. Sem. Abh. 5 (1927), 161-183.

46. Serre, J.-P., A Course in Arithmetic, Graduate Texts in Mathematics 7, Berlin, Springer-Verlag, 1973.

47. Shelah, S., "Infinite abelian groups, Whitehead problem and some constructions", Israel J. Math. 18 (1974), 243-256.

48. _________, "The lazy model theorist's guide to stability", to appear in the proceedings of a conference on model theory, Louvain-la-Neuve, Belgium, 1975.

49. _________, Stability and the Number of Non-isomorphic Models, Amsterdam, North-Holland, to appear.

50. Shoenfield, J., Mathematical Logic, Appendix, Reading, Mass., Addison-Wesley, 1967.

51. Simmons, H., "An omitting types theorem with an application to the construction of generic structures", Math. Scand. 33 (1973), 46-54.

52. Szmielew, W., "Elementary properties of abelian groups", Fund. Math. 41 (1955), 203-271.

53. Warfield, R., "Purity and algebraic compactness for modules", Pac. J. Math. 28 (1969), 699-719.

54. Wheeler, W., Algebraically Closed Division Rings, Forcing and the Analytic Hierarchy, dissertation, New Haven, Yale University, 1972.

55. Fried, M. and Sacerdote, G., "A primitive recursive decision procedure for the theory of almost all finite fields and all finite fields", to appear.

56. Cherlin, G., "Algebraically closed commutative rings", J. Symb. Logic 38 (1973), 493-499.

57. Lipshitz, L. and Saracino, D., "The model companion of the theory of commutative rings without nilpotent elements", PAMS 38 (1973).

58. Macintyre, A., "On ω_1-categorical theories of fields", Fund. Math. 71 (1971), 1-25.

59. Ribenboim, P., Theorie des Valuations, Univ. Montreal Press, Montreal (1964).

60. Kaplansky, I., "Maximal fields with valuations", Duke Math J. 9 (1942), 303-321.

Subject Index

Index of Principal Notation

Page	Notation	Meaning
108	DR	Division rings
114	$h(x)$	Height of term x
120	rk	Rank
124	$\langle\langle X \rangle\rangle$	Division ring generated by X
125	$C(x)$	Centralizer
125	$C^2(X)$	$C(C(X))$
133	^1N	First order arithmetic
133	^2N	Second order arithmetic
134	$^1N(D), {}^2N(D)$	Models of arithmetic in D
156	M_R	R-modules
156	AG	Abelian groups
159	$M_1 \oplus_M M_2$	Amalgamated sum
186	Z/p^∞	Prüfer p-group
186	Z_p	Z localized at p
186	$\lvert a \rvert_p$	p-norm
188	$\lvert a \rvert$	$\sum 2^{-p} \lvert a \rvert_p$
189	$m \mid x$	m divides x
195	$\alpha, \beta, \gamma, \delta$	Szmielew invariants
215	Tr	Trace
217	$t(x)$	$x^p - x$